To Mayo, Fabián, Christián and Iván.

CONTENTS

PREFACE

The advances of Microelectronics have made possible the realization of high-performance electronic systems. They require the design of very complex electronic circuits. Complex circuitry very often implies the implementation of mixed-mode systems, that is analog and digital mixed in the same system. Two of the main blocks that make the link between the analog and digital circuitry are the analog-digital converters and analog filters. If cost, reliability, performance, size and power consumption are of primary interest, then these blocks must be realized as integrated circuits.

The analog-digital converters are not discussed in this text. The filters which can be fabricated in MOS technologies are switched-capacitor and time continuous filters. Actually, the switched-capacitor filtering technique is mature and several excellent text books have already been published. A filtering technique that has recently emerged is the time continuous one. This text is devoted to the design of CMOS continuous-time filters. CMOS is employed because the most complex integrated circuits have been realized with this technology since two decades.

In this text, the most important advantages and drawbacks of the continuous-time filters are clearly shown. The transfer function is one of the most important filter parameters but several other ones (like intermodulation distortion, power-supply rejection ratio, noise level and dynamic range) are fundamental in the design of high-performance systems. Special attention is paid to the practical aspects of the design, which shows the difference between an academic design and an industrial design. A clear understanding of the behavior of the circuits and techniques is preferred over complex equations or interpretations of simulated results. Step by step design procedures are very often used to clarify the use of the techniques and topologies.

Design techniques for high-performance CMOS continuous-time filters can be used as a text book for senior or graduate courses on this topic. Also, this text is useful for industrial engineers as a reference book. The organization of the text is hierarchical, starting with the design considerations of the basic building blocks and ending with the design of several high-performance continuous-time filters. Most of the circuits have been fabricated, theoretical analyzed, simulated and silicon measurement results are compared with each other.

José Silva-Martínez
Autonomous University of Puebla

Michiel Steyaert
Willy M. C. Sansen
Catholic University Leuven

CHAPTER

1

INTRODUCTION

1.1. Analog IC filters.

The actual trend of Integrated Circuits (IC) are both to scale down the dimensions of the transistors and to incorporate in a single chip as many building blocks as possible. In the design of high-performance electronic circuits the use of analog filters is unavoidable. The advance of the Very-Large Scale Integration (VLSI) techniques demands for high-frequency and high-performance active filtering. In Telecommunication applications, for instance, active filters ranging from a few kiloHertz (kHz) up to several MegaHertz (MHz) are required. Some other communication systems like radio and video frequency receivers demand high-frequency and high-selectivity analog filters as well.

The very small silicon area required by the digital circuits, their extremely large dynamic range, their very low cost and efficient digital memory make the digital signal processors very attractive for ICs. Actually, available digital design tools can take a logic schematic and automatically convert this to a chip layout. However, these digital signal processors have to be interfaced with the real and in most cases analog world. For analog-digital interfaces, the use of high-frequency and high-dynamic range active filters is mandatory. Typically, these specifications can only be achieved with IC realizations.

Typically, the most important characteristics of the time continuous IC filter are sensitive to the process parameter tolerances. Actually, the available technologies for integrated circuits are Complementary-Metal-Oxide-Semiconductor (CMOS), Bipolar, BICMOS (BIpolar-CMOS), Gallium-Arsenide and devices based on III-V Compounds. Each one of these technologies have advantages and drawbacks. CMOS is, actually, the most popular technology due to its higher circuit density, low cost and maturity. Since at this time the mostly used techniques for the digital signal processing are implemented in CMOS technologies the design techniques are discussed for this technology. Nevertheless, most of these techniques can be used at least in Bipolar and BICMOS technologies. Also, only continuous-time filters that can be incorporated in digital systems are further discussed.

1.2. Full CMOS analog filters.

In CMOS technologies, the analog filter design techniques can be divided in analog sampled-data and time continuous techniques [1-8]. Switched-capacitor and switched-current filters can be grouped in the former technique [1,3-7]. In the last group, RC active filters, MOSFET-C filters and OTA-C filters can be associated [3, 8-15].

1.2.1. Sampled-data filters. These filtering techniques use several non-overlapping clock phases. In switched capacitor filters, the main characteristics are determined by a clock frequency and by capacitor ratios. In CMOS technologies both parameters can be controlled with an accuracy as high as 0.5% [3-6]. Even more, both parameters, clock frequency and capacitor ratios, are almost independent of the process parameter tolerances and temperature variations. Hence, a major advantage of this technique is the high accuracy of its integrator time constant. Therefore, the use of additional tuning circuits is avoided. These factors make these techniques very attractive for the design of high-performance analog integrated filters. In low-frequency applications, the high dc gain and high Gain-Bandwidth Product (GBW) of the CMOS OPerational AMPlifiers (OPAMPs) make these filters insensitive to the parasitic capacitors. Furthermore, as the poly-silicon capacitors are quite linear and due to the OPAMP local feedback (small ac-signal at its input terminals) the Total Harmonic Distortion (THD) of the switched-capacitor filters is very low, e.g. THD<-70 dB. Unfortunately, these advantages of switched-capacitor filters are not necessarily maintained for high-frequency applications. This is mainly due to the finite parameters of the OPAMP (finite dc gain and finite GBW), finite resistance of the switches and clock feedthrough [5,6]. In high-frequency applications the OPAMP has to be fast enough to settle to the right output within a half clock period. For a settling precision of 0.1 %, the settling time should be higher than the GBW of the OPAMP at least by a factor 7. However, due to the additional capacitors connected to the OPAMP output the effective settling time increases and as a result even larger GBWs are required. Typically a factor 10 (or more) instead of 7 is very often used [5]. In order to guarantee the stability of the closed loop system, the second pole of the OPAMP should be placed around 3 times higher than GBW. For high-frequency applications it is very difficult to satisfy last constraint because it means that the frequency of the OPAMP second pole should be higher than the clock frequency by a factor of 30. The other major limitation of switched-capacitor systems is the relatively low dc gain of the CMOS OPAMP. The dc gain of the OPAMP has to be high enough to reduce the ac-signal at its input terminals. If so, the circuit becomes parasitic insensitive and allows the total charge and discharge of the capacitors. Otherwise the precision of the filter is further reduced and the harmonic distortion increases rapidly [9].

The switched-current technique has recently been introduced [7, 8]. The principle of these filters is the same as the switched-capacitor filters principle but instead of charge packages it drives current packages. Thus, the transfer function depends of both clock frequency and current source ratios (transistor transconductance ratios). Since the accuracy of capacitor ratios is higher than the accuracy of transconductance ratios the precision of switched-current filters is lower than this of switched-capacitor filters [1-4]. If the current mirrors are properly designed, accuracy's in the order of 1% can be achieved. Switched-current filters can be implemented using transistors and the gate-channel capacitance of the CMOS transistors so avoiding the use of additional poly-silicon capacitors. Therefore this technique can be used in purely digital technologies. In principle, due to the absence of OPAMPs this type of filters can be faster than the switched-capacitor filters. Nevertheless, there are several practical limitations in these structures. A major disadvantage of this technique is the clock feedthrough. The clock feedthrough reduces the accuracy of the filter and increases both noise level and distortion [8]. In switched-current filters, due to the absence of large capacitors this effect is much more severe than in switched-capacitor filters. Another less critical disadvantage of this technique is that the current efficiency (ac current/bias current) decreases further for low-distortion applications [8]. Finally, this type of networks requires a linear voltage-to-current transducer at the input of the filter.

It should be mentioned that sampled-data filters need an antialiasing continuous-time filter to band-limit the frequencies of the input signal. However, in the case of purely analog signal processing a smoothing continuous-time filter has to be used as well. Due to the sampling, high-frequency noise can be aliased on to the base band increasing further the in-band noise level. This effect can drastically reduce the filter dynamic range. Lastly, it should be recognized that in some applications wherein several information channels must be separated it is not possible to use sampled data filters. This is mainly due to the aliasing effect. A filtering technique that avoids several of the above mentioned disadvantages is the continuous-time filtering technique.

1.2.2. Continuous-time filters. In principle, these filters were developed as complementary of switched-capacitor filters as antialiasing and smoothing filters. Nowadays, time continuous filters are an alternative to sampled-data filters in low-frequency applications [11-16]. Moreover, they allow the integration of filters in the MHz and several hundreds of MHz frequency ranges [17-26]. This technique avoids the need of pre- and post-filtering that in most of the cases switched-capacitor filters require. Because it is a time continuous technique, the aliasing problems are not present. The precision of these filters is in the order of 30 - 50 %. This is the major disadvantage of continuous-time filters. In order to improve this precision it is necessary to include an on-chip self tuning system in the system. This last fact complicates the design of the filter.

For the design of high-performance CMOS active filters, there are three main types of continuous-time filters, namely, RC active filters, MOSFET-C filters and OTA based filters. The accuracy of the active filters is very limited, e.g. ± 50 %. A technique that allows to improve the precision of these filters has been recently proposed [12]. This technique uses discrete capacitor banks and so its accuracy is limited by the resolution of the banks, e.g. accuracy ≈ ± 5 %. This technique however is not discussed in this text.

The MOSFET-C filters are basically the active RC filters but with the resistors implemented by equivalent CMOS tunable resistors [5,10,11]. Due to the use of the resistors, buffered OPAMPs have to be employed. The limited frequency response of the two or more stages OPAMPs reduce the use of those filters to low-frequency applications, e.g. less than 100 kHz.

OTA-R-C filters. The Operational Transconductance Amplifier-Capacitor (OTA-C) filters have already been reported for frequency ranges from a few kHz up to Very High Frequencies, 250 MHz, [13-26]. The basic building blocks of OTA-C filters are the Operational Transconductance Amplifier (OTA) and the poly-silicon Capacitors (C). Recently, linear full CMOS Resistors (R) have also been incorporated in filters leading to the OTA-R-C filters [20,24-27]. The monolithic filter realizations are commonly based on lossless integrators. Two basic lossless continuous-time integrators are shown in Fig. 1.1.

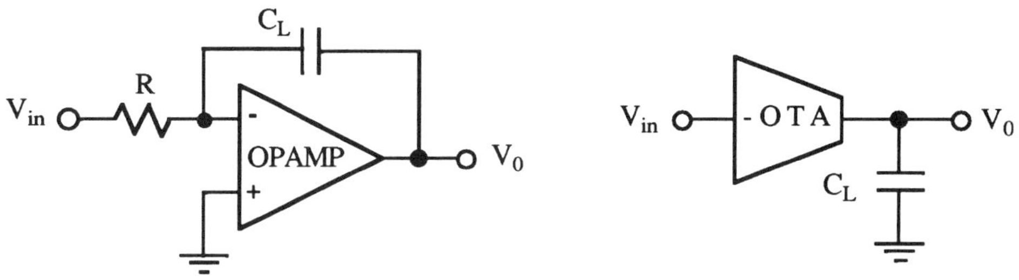

(a) $\frac{V_0}{V_{in}}(s) = -\frac{1}{s\,R\,C_L}$ (b) $\frac{V_0}{V_{in}}(s) = -\frac{g_m}{s\,C_L}$

Fig. 1.1. Continuous-time lossless integrators:
a) RC active and b) OTA-C.

For the RC active integrator, the integrating constant is determined by the RC_L product. For the OTA-C realization this time constant is given by C_L/g_m, with g_m the transconductance of the OTA. C_L/g_m depends of the transistor and capacitor dimensions, transistors operation point and process parameters. The process parameters used in this text are given in appendix A1. According to Fig. 1.1 it is clear that the OTA transconductance plays a similar role in OTA-C filters as the resistors in RC active filters. Hence, the whole RC active filter design technique can be used for the design of OTA-C filters.

The OTA is basically a voltage-to-current transducer and the capacitor is a typical poly-silicon capacitor. In the ideal case, the capacitor is a noiseless device and so the noise level of the integrator is determined by the OTA. Typically, the signal to noise ratio of the OTA-C filters is in the range of 60 to 90 dB. This dynamic range is slightly lower but comparable to the dynamic range of the RC active and switched-capacitor realizations. Unfortunately, the total dynamic range of the OTA-C filters can not be fully exploited. A major drawback is the lack of very linear and efficient voltage-to-current transducers. Due to the relatively high distortion of the first generation of voltage-to-current transducers, the harmonic distortion components of the earlier designs were in the order of -40 dB for input signals up to 1 Volt [13-16]. More recently, the use of linearized topologies has increased the signal to harmonic distortion ratio of the continuous-time filters to the range of 52 dB up to 65 dB [16-20,24-26]. Typically, the signal to harmonic distortion components ratio for the low-frequency switched-capacitor filters are in the range of 70 to 90 dB. Obviously, the relatively higher harmonic distortion of continuous-time filters reduces their range of applications. Other major drawback of continuous-time filters is the necessity of an additional automatic tuning system.

A major advantage of the OTA based continuous-time filters is their extremely large frequency response [16-23,26]. Actually, OTA based filters with resonant frequencies in the range of several hundreds of MHz have been reported [22-23]. A comparison of the useful frequency range for both OPAMP and OTA based circuits is shown in Fig. 1.2.

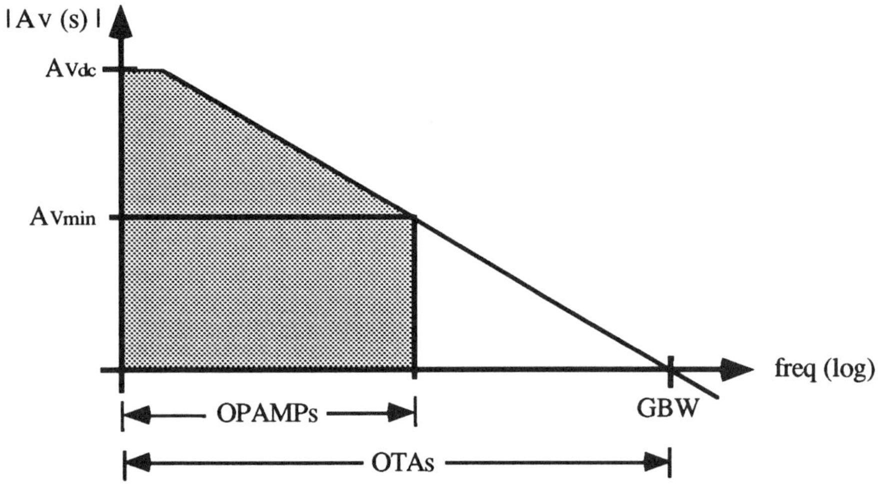

Fig. 1.2. Frequency response for a typical single pole OPAMP/OTA.

A minimum OPAMP gain, around 30 dB, is needed for well behaved OPAMP based circuits. Otherwise the GBW effects of the OPAMP degrades the characteristics of the system. For a single pole OPAMP this implies a useful frequency range of the order of GBW/30. This fact makes it extremely difficult to design high-frequency OPAMP based filters. Because the OTA based integrator is implemented in open loop (see Fig. 1.1) the useful frequency range of these circuits is limited only by the OTA gain-bandwidth product.

1.3. Description of the text.

This text deals with the analysis and design of full CMOS high-performance OTA based filters. The design details for the building blocks (OTA, CMOS resistors and capacitors) are further considered. Frequency range of operation, harmonic distortion components, signal to noise ratio, rejection to supply noise and filter precision are of primary concern in the design of the filters. Several design techniques and practical considerations for the optimization of these parameters are discussed. Special attention is paid to the fact that these designs can be incorporated in typical 5 Volt digital systems. The feasibility of the design techniques is shown by the experimental results of three filters. The use of low-distortion voltage-to-current transducers, linear CMOS resistors and an automatic tuning system results in signal to noise ratios in the order of 60-65 dB. These achievements enable the use of continuous-time filters for several high-performance applications.

The organization of the text is hierarchical: from the device level up to the system level. Firstly, the major building blocks of OTA-R-C filters, OTAs and CMOS resistors and capacitors, are studied (devices level). Afterwards, the biquadratic filters and the automatic tuning systems are discussed (building blocks level). The design of the higher order filters, lowpass and bandpass, are considered in the last part of the text (system level).

The major design considerations for the core of the filter, the OTA, are discussed in chapter 2. Several techniques for the design of high-performance OTAs, single-ended and fully-differential, are discussed [24-26]. These techniques allow to design both low and high-frequency continuous-time filters with very low harmonic distortions, e.g. total harmonic distortions lower than -60 dB. In the same chapter, an efficient technique for the control of the Common-Mode FeedBack (CMFB) of the fully-differential OTAs is discussed [25]. This technique reduces both power consumption and silicon area.

In chapter 3, the effects of the parasitic capacitors and the implementation of several CMOS resistors are considered. The parasitic capacitors affect the absolute value of the integrator time constant. As a result, it will decrease the accuracy of the filter. Because the continuous-time filters are sensitive to these parasitic capacitors, they have to be included in the system calculations. Moreover, the junction capacitors increase the total harmonic distortion of the system. In order to reduce the harmonic distortion due to these capacitors, some design considerations have to be taken into account. These design considerations are discussed as well. In the same chapter, several grounded and floating full CMOS resistors are described [24-26]. The major advantages of these realizations are both high linearity and high tunability range. Due to the larger tunability range of the full CMOS resistors compared to the typical OTA based resistors, the programmability of the OTA-R-C filters is much higher. These resistors are used in the realizations of several high-performance continuous-time filters.

In chapter 4, the problem of the automatic tuning of continuous-time filters is addressed. A major disadvantage of continuous-time filters is their low accuracy. The most important reasons for this low accuracy are described in this chapter. A comparison of the most commonly used second-order continuous-time filters is carried out. The major characteristics of these topologies are further discussed, e.g. higher programmability of the typical biquadratic filters and lower passband sensitivity of the ladder filters. The effects of the OTA finite parameters on the filter performance are studied as well. In order to overcome the shortcomings of continuous-time filters, an on-chip automatic tuning system has to be included in the design. The typical PLL based systems are discussed in the same chapter [13,14,17-21]. In addition, a different approach based on both a charge comparison principle and time domain analysis is described [26,27]. These techniques allow to reduce the tuning signal feedthrough of the system. As a result, the dynamic range of the filter is little sensitive to the operation of the automatic tuning system.

In chapter 5, the described techniques are used in the design of several high-performance filter realizations. The design and characterization of an equiripple fourth-order fully-differential 100 kHz is described. The results for this filter have shown a filter dynamic range in the order of 66 dB while the total harmonic distortion is lower than -60 dB. In the same chapter, two bandpass filters are discussed. A 1.8 MHz bandpass filter with programmable quality factor over more than 1 decade, from 2 up to 50, is introduced. The extremely large programmability range is a result of the large tunability range of the CMOS resistor employed. The last design is a 10.7 MHz fourth-order bandpass filter with a bandwidth of 300 kHz and on-chip automatic tuning system [26]. The experimental results for the filter, including the automatic tuning system, have shown a dynamic range of 68 dB with the input referred noise integrated over the filter passband. The design procedure of both filter and automatic tuning system is well described in this chapter.

In the last chapter of the text general conclusions are given.

REFERENCES

[1] P. R. Gray, D. A. Hodges and R. W. Brodersen, editors, *Analog MOS Integrated Circuits*, IEEE PRESS, New York:NY, 1980.

[2] P. E. Allen and D. R. Holberg, *CMOS Analog Circuit Design*, Holt, Rinehart and Winston, New York: NY, 1987.

[3] R. Gregorian and G. C. Temes, *Analog MOS Integrated Circuits for Signal Processing*, John Wiley & Sons, New York:NY, 1986.

[4] D. J. Allstot and W. C. Black, "Technological Considerations for Monolithic MOS Switched-Capacitor Filtering Systems," Proc. of the IEEE, Vol. 71, No 8, pp. 967-986, Aug. 1983.

[5] Y. Tsividis and P. Antognetti, *"Design of MOS VLSI Circuits for Telecommunications, editors,"* Prentice-Hall, inc., New Jersey, 1985.

[6] B. S. Song, "A 10.7 MHz Switched-Capacitor Bandpass Filter," IEEE J. of Solid-Sate Circuits, Vol. 24, pp. 320-324, April 1989.

[7] J. B. Hughes, N. C. Bird and I. Macbeth, "Switched Currents: A New Technique for Analog Sampled Data Processing," IEEE Proc. ISCAS-89, Portland, pp. 1584-1587, May 1989.

[8] R. T. Baird, T. S. Fiez and D. J. Allstot, "Speed and Accuracy Considerations in Switched-Current Circuits," IEEE Proc. ISCAS-91, Singapore, pp. 1809-1812, June 1991.

[9] F. Op 't Eynde, "High Performance Analog Interfaces for Digital Signal processors," Ph. D. Thesis, Katholieke Universiteit Leuven, Nov. 1990.

[10] A. S. Sedra and P.O Brackett, *Filter Theory and Design: Active and Passive*, Matrix Publishers, Inc., Champaign Illinois, 1978.

[11] M. Banu and Y. Tsividis, "Fully integrated active RC filters in MOS technology," IEEE J. of Solid-State Circuits, Vol. SC-18, pp. 644-651, Dec. 1983.

[12] J. B. Hughes, N. C. Bird and R.S. Soin, "A Novel Digitally Self-Tuned Continuous-Time Filter technique," IEEE-ISCAS-86, San Jose, pp. 1177-1180, may 1986.

[13] K. Tan and P. Gray, "Fully Integrated Analog Filters Using Bipolar-JFET Technology," IEEE J. of solid-State Circuits, Vol. SC-13, pp. 814-821, Dec. 1978.

[14] K. Moulding, J. Quartly, P. Rankin, R. Thompson and G. Wilson, "Gyrator Video Filter IC with Automatic Tuning," IEEE J. of Solid-State Circuits, Vol. SC-15, pp. 963-967, Dec. 1980.

[15] R. L. Geiger and E. Sanchez-Sinencio, "Active Filter Design Using Operational Transconductance Amplifiers: A Tutorial," IEEE Circuits and Devices Magazine, Vol. 1, pp. 20 - 32, March 1985.

[16] H. Khorramabadi and P. R. Gray," High Frequency CMOS Continuous-Time Filters," IEEE J. of Solid-State Circuits, Vol. SC-19, pp 939-948, Dec. 1984.

[17] C. S. Park and R. Schawmann, " Design of a 4-MHz Analog Integrated CMOS Transconductance-C Bandpass Filter," IEEE J. of Solid-State Circuits, Vol. SC-23, pp. 987-996, Aug. 1988.

[18] F. Krummenacher and N. Joehl," A 4 MHz CMOS Continuous-Time Filter with On-Chip Automatic Tuning," IEEE J. of Solid-State Circuits, Vol. SC-23, pp 750-757, June 1988.

[19] V. Gopinathan and Y. P. Tsividis," A 5V 7th-Order Elliptic Analog Filter for Digital Video Applications," IEEE Int. Solid-State Circuits Conference, San Francisco Calif., pp 208-209, Feb. 1990.

[20] Y. T. Wang, F. Lu and A. A. Abidi,"A 12.5 MHz CMOS Continuous Time Bandpass Filter," IEEE Int. Solid-State Circuits Conference, New York, pp 198-199; Feb. 1989.

[21] J. M. Khoury, "A 15 MHz Continuous-Time Bessel Filter for Disk Drives," IEEE Intl. Solid-State Circuits Conference, New York, pp. 134-135, Feb. 1991.

[22] M. Snelgrove and A. Shoval, “A CMOS Biquad at VHF,” IEEE Proc. CICC-91, San Diego, pp. 9.2.1-9.2.6, May 1991.

[23] B. Nauta, “CMOS VHF Transconductance-C Lowpass Filter,” IEE Electronics Letters, Vol. 26, pp. 421-422, March 1990.

[24] J. Silva-Martinez, M. Steyaert and W. Sansen, "A Large-Signal Very Low-Distortion Transconductor for High-Frequency Continuous-Time Applications," IEEE J. of Solid-State Circuits, Vol. SC-26, pp. 946-955, July 1991.

[25] J. Silva-Martinez, M. Steyaert and W. Sansen, "Design Techniques for High-Performance Full CMOS OTA-R-C Continuous-Time Filters," IEEE J. of Solid-State Circuits, pp. 993-1001, July 1992.

[26] M. Steyaert and J. Silva-Martinez, "A 10.7 MHz CMOS OTA-R-C Bandpass Filter with 68 dB Dynamic Range and On-chip Automatic Tuning," IEEE Int. Solid-State Circuits Conference, San Francisco, pp. 66-67, Feb. 1992.

[27] J. Silva-Martinez, M. Steyaert and W. Sansen, "A Novel Approach for the Automatic Tuning of Continuous-Time Filters," IEEE Proc. ISCAS-91, Singapore, pp. 1452-1455, June 1991.

Appendix A1
The process parameters

In this appendix, the process parameters of the used technologies are given. In table A1.1, the parameters for the 3 μm double poly double metal N-well CMOS process are shown. In table A1.2, the double metal CMOS 1.5 μm N-well process parameters are depicted. Finally, in table A1.3, the process parameters of the 1.2 μm double metal CMOS N-well technology are given.

parameter	NMOS transistor	PMOS transistor	units
level	2	2	
V_T	0.9	- 0.9	V
K_P	57 E-6	17 E-6	A/V
t_{OX}	425 E-10	425 E-10	meters
GAMMA (γ)	0.3	0.5	$V^{-1/2}$
PHI (ϕ)	0.7	0.69	V
LAMBDA (λ)	0.02 (L=9μm)	0.01 (L=9μm)	V^{-1}
LD	0.22	0.135	μm
DL	0	0	μm
DW	-0.4	- 0.4	μm
CGS0	1.76 E-10	2.8 E-10	F/m
CGD0	1.76 E-10	2.8 E-10	F/m
C_J	0.7 E-4	3.3 E-4	F/m^2
M_J	0.5	0.5	-
C_{JSW}	3.9 E-10	4.4 E-10	F/m
M_{JSW}	0.33	0.33	-
J_S	1 E-3	1 E-3	A/m^2
RSH	25	45	Ω/Sq
U_{CRIT}	1 E4	1 E4	V/cm
NFS	1 E11	1 E11	cm^{-2}

Table A1.1. Spice parameters for the 3 μm double poly double metal CMOS N-well process.

parameter	NMOS transistor	PMOS transistor	units
level	3	3	-
V_T	0.783	- 1.016	V
K_P	68.1 E-6	23.8 E-6	A/V
t_{OX}	280 E-10	288 E-10	meters
N_{SUB}	1.264 E16	1.949 E16	$1/cm^{-3}$
GAMMA (γ)	0.525	0.67	$V^{-1/2}$
PHI (ϕ)	0.465	0.863	V
U0	554	194.86	$cm^2/V.S$
THETA (θ)	0.046	0.120	1/V
DELTA	2.481	1.069	-
LAMBDA (λ)	0.02 (L=9μm)	0.01 (L=9μm)	V^{-1}
LD	0.3	0.3	μm
DL	0.09	0.162	μm
DW	-0.204	- 0.279	μm
CGS0	3.1 E-10	3.1 E-10	F/m
CGD0	3.1 E-10	3.1 E-10	F/m
C_J	2.56 E-4	4.32 E-4	F/m^2
M_J	0.404	0.5	-
C_{JSW}	2.32 E-10	2.78 E-10	F/m
M_{JSW}	0.379	0.395	-
J_S	1 E-3	1 E-3	A/m^2
RSH	34.3	82.2	Ω/Sq
U_{CRIT}	1 E4	1 E4	V/cm
NFS	1 E11	1 E11	cm^{-2}
PB	0.708	0.829	V
ETA (W/L=30/3)	0.016	0.014	-
KAPPA	0.43	7.99	-
V_{MAX}	1.2 E5	2.12 E5	m/s

Table A1.2. Spice parameters for the 1.5 μm double metal CMOS N-well process.

parameter	NMOS transistor	PMOS transistor	units
level	3	3	-
V_T	0.715	- 0.784	V
K_P	78.2 E-6	20.5 E-6	A/V
t_{OX}	200 E-10	208 E-10	meters
N_{SUB}	2.944 E16	2.45 E16	1/cm^{-3}
GAMMA (γ)	0.604	0.543	$V^{-1/2}$
PHI (ϕ)	0.462	0.9	V
U0	507	170.2	$cm^2/V.S$
THETA (θ)	0.0636	0.181	1/V
DELTA	1.1	2.57	-
LAMBDA (λ)	0.02	0.01	V^{-1}
LD	0.21	0.21	μm
DL	0.3	0.3	μm
DW	-0.204	- 0.279	μm
CGS0	3.31 E-10	3.41 E-10	F/m
CGD0	3.131 E-10	3.41 E-10	F/m
C_J	3.0 E-4	5.4 E-4	F/m^2
M_J	0.384	0.44	-
C_{JSW}	3.31 E-10	5.7 E-10	F/m
M_{JSW}	0.247	0.34	-
J_S	1 E-3	1 E-3	A/m^2
RSH	60	112	Ω/Sq
U_{CRIT}	1 E4	1 E4	V/cm
NFS	1 E11	1 E11	cm^{-2}
PB	0.6	0.6	V
ETA	0.07	0.05	-
KAPPA	0.1	1.6	-
V_{MAX}	1.5 E5	1.2 E5	m/s

Table A1.3. Spice parameters for the 1.2 μm double metal CMOS N-well process.

CHAPTER

2 VERY LOW-DISTORTION OPERATIONAL TRANSCONDUCTANCE AMPLIFIERS FOR HIGH-FREQUENCY APPLICATIONS

2.1. Introduction.

In the previous chapter, it has been mentioned that the characteristics of analog filters are determined by the precision of both the integrator time constant and the dc gain of the lossy integrators [1]. Several techniques for the implementation of full CMOS continuous-time integrators have been already reported [2-5,7,12-22,24,25]. These design techniques can be divided in RC active filters, MOSFET-C filters and OTA-C filters.

The RC active filters are still being used in low-frequency applications. These filters use OPAMPs, resistors and capacitors as basic elements. In integrated circuits, the resistors can be implemented either as diffused or poly-silicon resistors. The absolute accuracy of the resistance is in the order of 30% [6,8]. The linearity of the poly-silicon resistors is quite good and the accuracy of resistor ratios can be as good as 1%. Nevertheless, the accuracy of the RC products can be as worse as 50%. Even more, the RC product is a strong function of both technology tolerances and temperature variations. In filtering applications, the frequency of the poles and the frequency of the zeros is determined by RC products. Typically, the quality factor of the filter is determined by resistors and/or capacitor ratios. Because the filters are more sensitive to variations in the frequency of the poles than to the quality factor of the poles, the accuracy of the active RC filters is quite low. The accuracy of these filters can be improved if certain kind of tuning, either on chip or externally, is included. The active RC filter approach is almost not used for high-frequency applications because there are some other major disadvantages; e.g. the resistors are implemented in long strips over field oxide or over active area. So there are distributed capacitors connected to the resistor. These distributed capacitors limit both the precision and the high-frequency performance of the filter. Furthermore, due to the use of the resistors buffered OPAMPs are required. The two poles buffered OPAMP limits the frequency response of the system and increases dramatically the power consumption. Nevertheless, low-frequency filters with dynamic range higher than 75 dB have been reported [3].

The MOSFET-C filters are implemented with OPAMPs, CMOS transistors biased in the linear region and capacitors [2, 7]. These topologies are the same as the typical RC active filters but with the resistors are implemented by a MOS transistor biased in its linear region. In order to diminish the distortion, sometimes a transistor array instead of a single transistor is used [20]. Due to the simulated resistors, low output impedance OPAMPs must be used. For the circuit designer that means that OPAMPs are needed with at least two stages. As in the case of active RC filters, the limited frequency response of the two stages OPAMP and the distributed gate capacitance of the MOS transistor biased in linear region reduces their use to low-frequency applications, e.g. less than 100 kHz. In fact, the major applications of these filters are at low frequencies in the design of antialiasing and smoothing filters [2,7].

A more versatile technique is the so-called OTA-C. OTA-C filters have already been reported for frequency ranges from audio frequencies (a few kHz) up to very high-frequency (200 MHz) [4, 5, 11-22, 24, 25, 30]. The discussion of the practical limitations of the major building block of OTA-C filters, the OTA, is one of the subjects of this chapter. The OTA is a voltage-to-current transducer with very high output impedance. The C stands for a capacitor implemented in CMOS technology.

The resolution of OTA based circuits, and in general in circuits, is limited by the noise or unwanted signals. In the system, three types of noise or unwanted signal sources are present. Firstly, we have the signals generated by the intrinsic non-linearities of the electronic devices itself. Due to the non-linear nature of the relationship between the input signals applied to the devices and the output signals, several unwanted signals are generated. In the case of a simple sinusoidal input these signals are at multiple frequencies of the incoming signal frequency. These signals are called the harmonic distortions of the output signal [8]. This kind of noise is accounted by the Total Harmonic Distortion (THD). The THD is the RMS value of the contribution of all the harmonic distortions. If two or more signals with different frequencies are present at the input, sum frequencies and difference frequencies are generated at the output of the circuit although. These unwanted signals lead to the intermodulation distortions, to be defined shortly. Both THD and intermodulation distortion are the main limitations of the systems for large signals.

The second type of noise is due to the coupling of the device to the surrounding circuitry. The coupling is due to the finite impedance of the power-supply sources and to the coupling capacitors in the chip, namely: bulk capacitors, metal-metal capacitors, etc. As a result, signals generated in other part of the chip interferes with the filter signals. The quality of a system to be immune to this kind of noise is measured by the common-mode rejection ratio and power-supply rejection ratio parameters. For high-performance applications, special attention must be paid to these parameters.

The third type of noise is due to the noise generated in the device itself. This noise is due to both the discrete nature of the current flowing through the transistors and also due to the imperfections of the transistor interfaces, e.g. silicon-silicon oxide interface. In this chapter, the different type of noise sources present in an OTA are further discussed.

Because it is intended to incorporate these systems in mixed-mode systems (analog circuitry and digital circuitry are integrated on the same chip) the power-supply is only $\pm$ 2.5 Volts. However, because the trend of integrated circuits is to scale down both transistor dimensions and supply voltages, only those techniques that can be compatible with this trend are discussed. Nevertheless, some other interesting techniques can be found in the references listed.

The chapter is organized as follows. The characterization and the effects of the OTA non-linearities are investigated in section 2.2. Also, some of the linearization techniques published in the literature are studied. The study results in a Low-Distortion OTA (LDOTA) for high-frequency continuous-time applications. Its THD is lower than -60 dB (resolution of 10 bits) for input signals up to 2.4 Volts peak to peak (Vpp). The LDOTA second-order effects and its frequency response limitations are discussed in section 2.3. In the same section it is shown that the bulk effects are a major limitation of the topology for ultra-low distortion applications. The rejection to supply noise and the noise performance of the LDOTA are discussed in section 2.4. The effects of transistor mismatches and those due to the differential-to-single ended conversion are also considered. A detailed analysis of the LDOTA noise factor is also carried out. In order to increase the dynamic range of the LDOTA and, also, in order to reduce further both even-order non-linearities and common-mode noise, a fully differential structure is presented in section 2.5. The benefits and drawbacks of fully-differential structures are discussed in the same section. Because differential structures require a Common-Mode Feedback System (CMFB), the second-order effects of the

CMFB like mismatches, non-linearities and limited bandwidth are considered. Finally, some conclusions are given in the last section.

2.2. A Very Low-Distortion High-Frequency OTA.

In general, the range of signals that can be accurately driven by the electronic devices is limited in amplitude and in frequency. For large signals, the linear range of the systems is in the best case limited by the supply voltages. Very often the non-linearities of the building blocks reduce drastically this limit. In this text, dynamic range is defined as the ratio of the maximum input signal (RMS value) driven by the device to ensure a Total Harmonic Distortion (THD) lower than a specified value to the total input referred noise, integrated over the frequency band of interest. Typically, in the specification of the dynamic range the THD is used as reference. In order to reduce the effects of the harmonic distortion components, it is necessary to clearly define what the origin is of the harmonic distortion components and what their effects are. This is the goal of the next section.

2.2.1 Effects of the OTA non-linearities. OTA based filters use voltage-to-current transducers and capacitors as the basic elements. In the OTA, the relationship between the output current and the input voltage is function of the OTA operation point, input voltage, output current, temperature and process parameters. This nonlinear function is called large-signal transconductance, $g_M = i_o/v_{in}$. The OTA output current, i_o, and the associated capacitor C determines the integrator output voltage v_o (see Fig. 1.1 of chapter 1) as

$$v_o(t) = v_o(t_o) + \frac{1}{C}\int_{t_o}^{t} i_o \, dt = v_o(t_o) + \frac{1}{C}\int_{t_o}^{t} g_M v_{in} \, dt \tag{2.1}$$

with v_{in} and $v_o(t_o)$ the OTA input voltage and the initial condition of the integrator, respectively. In the ideal case both parameters g_M and C are constants. Nevertheless, both parameters are voltage and frequency dependent functions. In this section, C is considered as a voltage independent capacitor. Its non-linearities are discussed in the next chapter. If the OTA large-signal transconductance g_M is considered as a nonlinear function, the OTA output current can be expanded in a Taylor series around the operation point as

$$i_o = I_{DC} + \left[\frac{\partial i_o}{\partial v_{in}}\right]_Q v_{in} + \frac{1}{2}\left[\frac{\partial^2 i_o}{\partial v_{in}^2}\right]_Q v_{in}^2 + \frac{1}{6}\left[\frac{\partial^3 i_o}{\partial v_{in}^3}\right]_Q v_{in}^3 + \ldots \tag{2.2}$$

where Q means evaluation at the operation point. In this equation, the first term represents the output dc current of the OTA. The coefficient of v_{in} corresponds to the small-signal transconductance of the OTA and is denoted by g_m. The next two terms are the second and third-order non-linearities, respectively. If a single sinusoidal input signal, $v_{in} = V\sin(\omega t)$, is assumed then the output current becomes

$$\begin{aligned} i_o = I_{DC} &+ \left[\left[\frac{\partial^2 i_o}{\partial v_{in}^2}\right]_Q + ..\right]\frac{V^2}{4} + \left[\left[\frac{\partial i_o}{\partial v_{in}}\right]_Q + \left[\frac{\partial^3 i_o}{\partial v_{in}^3}\right]_Q \frac{V^2}{8} + ..\right] V\sin(\omega t) + \\ &+ \left[-\frac{1}{4}\left[\frac{\partial^2 i_o}{\partial v_{in}^2}\right]_Q + ..\right] V^2 \sin(2\omega t) + \\ &+ \left[-\frac{1}{24}\left[\frac{\partial^3 i_o}{\partial v_{in}^3}\right]_Q + ..\right] V^3 \sin(3\omega t) + .. \end{aligned} \tag{2.3}$$

In this equation, it can be noted that the effects of the second-order non-linearity leaves a dc term and an harmonic component at 2ω. Both components are proportional to the squared amplitude of the input signal. For dc and very low-frequency applications, the effects of the dc component might be important. For instance in some telemetry applications, temperature sensors, pH meters, etc. For high-frequency applications this component can be important in magnitude and frequency modulators, but, commonly it can be neglected in filtering applications. The second-order non-linearities are characterized by the Second Harmonic Distortion (HD2). If a single frequency input signal is considered, HD2 is defined as the ratio of the amplitude of the signal component at 2ω to the amplitude of the signal at ω, neglecting the higher order non-linearities. Thus, from 2.3 it can be obtained

$$HD2 = \frac{\left[\dfrac{\partial^2 i_o}{\partial v_{in}^2}\right]_Q}{\left[\dfrac{\partial i_o}{\partial v_{in}}\right]_Q} \frac{V}{4} \tag{2.4}$$

It should be remarked that 1 / HD2 is the signal to noise ratio for this particular effect. The effects of the third-order non-linearity are twofold. Firstly, a signal at the same frequency as the incoming signal and proportional to V^3 is generated. This component is associated with the Third Intermodulation Distortion (IM3) to be defined shortly. It must be noted that the ideal output signal should be proportional to V. This harmonic distortion component makes to the OTA transconductance voltage dependent. Hence, for large amplitude signals, the OTA transconductance is affected by a factor proportional to V^2. The second effect of the third-order non-linearity is the generation of a frequency component at 3ω. This component is very often used for the characterization of IM3. It is easier to measure this component because a single input signal is needed and for measuring IM3 2 signals must be mixed. HD3 is defined as the ratio of the amplitude of the frequency component at 3ω to the amplitude of the signal at ω. According to expression 2.3, HD3 can be deducted as

$$HD3 = \frac{\left[\dfrac{\partial^3 i_o}{\partial v_{in}^3}\right]_Q}{\left[\dfrac{\partial i_o}{\partial v_{in}}\right]_Q} \frac{V^2}{24} \tag{2.5}$$

The third intermodulation distortion is measured using a two tone input signal with different frequencies, ω_1 and ω_2, and equal amplitude. In this case, IM3 is defined as the ratio of the harmonic at frequency $2\omega_1$-ω_2 to the amplitude of the input tones. However, IM3 is strongly related to HD3. It can be easily demonstrated that IM3 can be obtained from 2.3 as the ratio of the magnitude of the frequency component at ω due to the third-order non-linearity over the magnitude of the frequency component due to the incoming signal at ω. From relationships 2.3 and 2.5 IM3 becomes

$$IM3 = 3\ HD3 \tag{2.6}$$

The intermodulation distortion is quite important in practical systems, wherein not only a single input signal is present. In fact, in practical cases several mixed signals with different frequencies are present at the input of the system. So, the effects of the second-order non-linearities are the generation of unwanted signals at frequencies equal to the sum frequencies and difference frequencies of the incoming signals. The magnitude of the second harmonic distortion can be calculated by applying 2.4 to each pair of the incoming signals and adding their power and taking the square root of that power. Again, this effect can be very important in low-pass filters and in broadband applications. The same can be done for the third-order distortion except that instead of HD3 the third intermodulation distortion, IM3, must be used. The frequency of the components have to be calculated by the sum and difference of the frequency of one tone multiplied by a factor of 2 and the frequency of the second tone. Typically, these components are the most important ones in order to determine the signal to distortion ratio of the system. This is because most of these components fall in the passband of the system, even for narrow-band applications. A discussion of the nature of these distortions and methods for their reduction follows.

2.2.2. OTA linearization techniques. The relationship between the output current and the input voltage of the transistor is a nonlinear function. In the most simple case, it is an exponential function for the bipolar transistor and a quadratic function for the MOS transistor [2, 6, 8-10]. When second-order effects are included, these relationships become quite complex making the modeling of the transistors rather involved. As a result, the analysis of complex circuits requires computer simulations. The disadvantage of computer simulations is that it gives numbers and sometimes is quite difficult to get insight in the behavior of the circuit. In this text, instead of simulations simpler transistor models are used. When necessary, the complexity of the model is increased, e.g. when second-order effects are studied. In this approach, the accuracy of the design is sacrificed at the gain of expressions that can be easily interpreted by the designer are obtained. Anyway, at the end of the design carefully simulations have to be carried out in order to optimize the design. If better modeling of the circuit is needed, the simulations have to be strongly related to the physical layout.

In chapter 4, it will be shown that the active filters are very sensitive to the excess phase of the active blocks, mainly in narrow band applications [1]. Hence, the phase response of the voltage-to-current transducer is quite important. The most important facts in the design of voltage-to-current transducers for continuous-time filters are: linearity, phase response, noise level and power consumption. In addition, it is very desirable that the linearization technique should be lowly sensitive to transistor mismatches.

Source degenerated topologies. A major advantage of the voltage-to-current transducers based on a differential pair, is the fact that the even-order harmonic distortions are canceled in the ideal case. In fact, in the case of the differential pair (see Fig. 2.1a) the sum of the output currents i_{01} and i_{02} is forced to be equal to the bias current $2I_B$. Due to the symmetry of the circuit both output currents can be described by the same equation but interchanging $v_d = v_{i1} - v_{i2}$ by $- v_d$. If the output currents are expressed in a Taylor series as relationship 2.3, it is clear that the second-order non-linearity for both outputs is the same and with the same sign. Because the sum of these currents have to be a constant these terms vanish making HD2=0. In practice, due to both mismatches and finite output impedance of the transistors, a small HD2 is always present.

The case of the odd-order distortions is quite different. In order to reduce those distortions, linearization techniques have to be used. A technique that allows natural linearization due to its intrinsic feedback is the source degeneration technique (see Fig. 2.1b). Other important advantages of this structure are that it exhibits excess lead phase in its phase response and that this technique is lowly sensitive to transistor mismatches. Even more, the zero generated by the source degeneration can be used to compensate the excess lag phase introduced by the OTA parasitic poles [12,21].

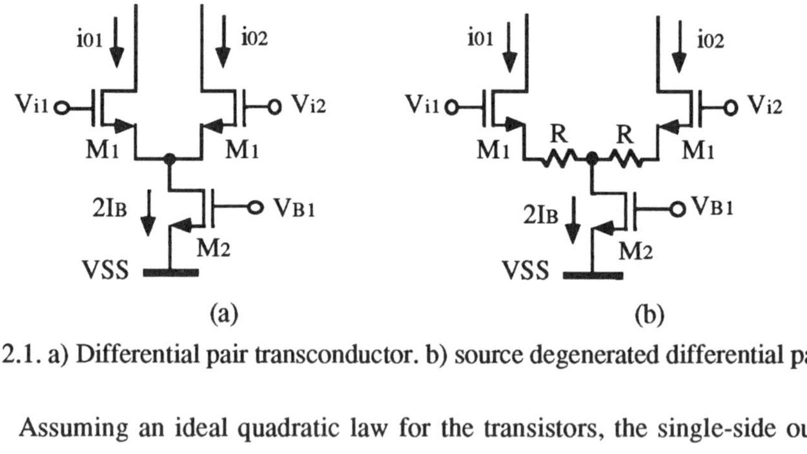

Fig. 2.1. a) Differential pair transconductor. b) source degenerated differential pair.

Assuming an ideal quadratic law for the transistors, the single-side output current of the source degenerated structures can be described by

$$i_{01} = I_B + \frac{\sqrt{2\beta_1 I_B}}{2} \frac{v_d}{n} \sqrt{1 - \left(\frac{v_d}{2n\left(V_{GS1} - V_T\right)}\right)^2} \tag{2.7}$$

with v_d the differential input voltage ($v_d = v_{i1} - v_{i2}$) and I_B and V_{GS1} the bias current and the gate-source bias voltage of M_1, respectively. V_T and β_1 are determined by technological and geometrical parameters [2, 8-10]. The n factor depends of the topology used. It is equal to 1 for the simple differential pair and approximately equal to $1/(1+g_m R)$ for the source degenerated topology. For a typical CMOS process the threshold voltage V_T is in the order of 0.7 Volts to 1 Volt. Hence, for supply voltages of ± 2.5 Volts, $V_{GS} - V_T$ should be smaller than 0.7 Volts in order to avoid the reduction of the common-mode range. For i_{02}, a similar equation to 2.7 can be written but interchanging v_d by $-v_d$. From 2.7 and 2.4 it can be easily demonstrated that the small-signal transconductance of the voltage-to-current transducer becomes

$$g_m = \left[\frac{\partial\left(i_{01} - i_{02}\right)}{\partial\left(v_1 - v_2\right)}\right]_Q = \frac{\sqrt{2\beta_1 I_B}}{n} \tag{2.8}$$

Similarly, it can be shown that HD2 vanishes. For HD3, mathematical manipulations lead to

$$HD3 = \frac{1}{32\, n^2}\left(\frac{V_d}{V_{GS1} - V_T}\right)^2 \tag{2.9}$$

For the case of the single differential pair, n = 1, and for $V_d = V_{GS1} - V_T$ the third-harmonic distortion is about -30 dB. The IM3 is in the order of -21 dB, which means a variation of the g_m of the order to 10% due the level of the input signal. For high-performance applications these distortions can not be tolerated. Certainly, the linearity can be improved by increasing the resistors of the source degenerated topology. Nevertheless, this approach requires larger resistors which increases both silicon area and parasitic capacitors. Those resistors have to be driven by larger transconductances, then the power consumption increases as well. However, the major drawback of the passive resistors is the lack of tunability. The larger the source degeneration the better the linearity but the lower the effect of the transistors is. In the limit, the transconductance of the source degenerated topology is equal to 1/R. As a result, the transconductor can not be tuned any more. For this reason, several CMOS realizations of tunable source degenerated topologies have been proposed [13-22,24].

The topology proposed by Krummenacher et.al. [13] is basically a source degenerated structure for small signals with an additional mechanism which increases its transconductance for large signals (see Fig. 2.2a). This mechanism is due to the finite transconductance of the transistors biased in linear region and the gate connection of the M_2 transistors. This fact increases the linear range of this topology in the order of a factor 2. For this topology, the factor n is given by $1+\beta_1/4\beta_2$. The best linearity of this topology is obtained for β_1/β_2 in the range of 6 to 10. If larger values are used, the effects of the differential pair reduce further leading to higher harmonic distortion components. Unfortunately, with this topology it is not possible to achieve very low harmonic distortions. For example, if $\beta_1/\beta_2 = 9$ and for a differential input voltage such that $V_d = V_{GS1}-V_T$, HD3 can be computed as -50.6 dB (see relationship 2.9). Actually, due to the additional current generated by the non-zero transconductance of transistors M_2 the distortion is in practice 2 or 3 dB lower leading to a HD3 ≈ -53 dB.

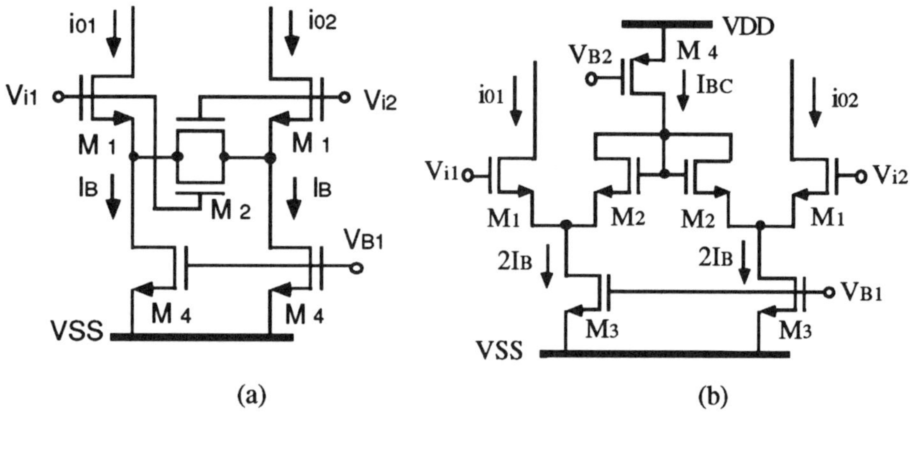

Fig. 2.2. CMOS source degenerated topologies using a) transistors biased in triode region and b) saturated transistors.

Another approach using source degeneration was proposed by Torrance et. al. [14]. In this technique, the resistors used for the source degeneration are simulated by saturated transistors with their drain connected to their gate. In this case, the factor n corresponds to the number of additional differential pairs used. This topology is shown in Fig. 2b for the case of 2 sections. In this technique, several differential pairs are connected in series such that the input voltage is split in the sections. Because the input voltage for each section is v_d / n, the transconductance is also reduced by the same factor n. The third-harmonic distortion, which is the most important one, is decreased by a factor n^2. Unfortunately, the phase response degrades rapidly for large number of sections, which limits this technique to 3 sections.

Other techniques. An interesting linearization technique was proposed by Nedungadi et. al. [15]. This technique uses two matched non-symmetrical differential pairs, or a feedback loop [16], in order to generate additional current for adjusting the bias current of the normal differential pair. Unfortunately, the non-symmetrical differential pairs increase the noise and the additional nodes are a limitation for the frequency response of the topology. Furthermore, due to the feedforward compensation this technique is very sensitive to transistors mismatches [17]. Techniques using class AB input stages [18] and also the ones using bias offset [19] very often require larger supply voltages. This is because in these techniques the transistors are connected in series, reducing further their common-mode range. Moreover, due to the additional nodes, their frequency response is limited. Other techniques using transistors biased in linear region and those using odd and even non-linearities cancellation improves the transconductor linearity. THD lower than -60 dB for input signals up to 2 Vpp and supply voltages of ±5 Volts have been already reported [2, 7]. A drawback of these techniques is that the noise increases [20]. Moreover, because the linearity depends on the gate voltage, the linearity is strongly reduced when low supply voltages are used. Nevertheless, recently a high-performance high-frequency filter using this technique has been reported [21]. The first high-frequency continuous-time filter reported in the literature was proposed by Khorramabadi et.al.[22]. In this filter, a cross-coupled technique was used. With this technique, the improvement in the linearity is limited. Even if the transistor mismatches are neglected, the third-harmonic distortion is higher than -50 dB for input signals lower than 1 Vpp. Some other techniques with higher harmonic distortions (THD > -40 dB) but useful for very-high-frequency applications have also been reported [36]. Because the main subject of this text is low distortion, these topologies are not discussed.

The harmonic distortion of the discussed techniques is lower than -55 dB for differential input signals with amplitudes lower than 1 Vpp. In the next section, a voltage-to-current transducer achieving THD < -60 dB for input signals up to 2 Vpp is discussed.

2.2.3. A low distortion voltage-to-current transducer. Regarding the good characteristics of the source degenerated structures, the design of the low distortion OTA is based on these techniques. So, it is guaranteed that HD2 is very low and we will try to reduce the HD3. In order to increase the linearity of the Fig. 2.2a structure, the splitting technique can be used as shown in Fig 2.3 [25]. Hence, the factor n increases by 2, $n = 2(1+\beta_1/4\beta_2)$. From expression 2.9 it can be easily seen that HD3 is reduced by a factor of 4 (12 dB). It should be clear that using this technique, the design of OTAs for high-frequency applications with HD3 lower than -60 dB is feasibly. In reality, in order to have this reduction in the harmonic distortion the same V_{GS} and the same transconductance have to be used. Then, the total bias current should be increased by a factor of 4 and the width of the transistors M_1-M_3 have to be increased by a factor of 2. Two disadvantages of this structure are the additional transistors and the increment of the power consumption. The first drawback is not really important but it represents some additional silicon area. The second shortcoming can be overcome using a cross coupled technique. If the transistors connected to the common-mode node, gate of M_3 in Fig. 2.3, are split in transistors M_3 and M_4, as shown in Fig. 2.4, and the in phase current components are added, the single-side output current becomes

$$i_{01} = I_{01} - K_C\left[\frac{\sqrt{2\beta_1 I_B}}{2}\,\frac{v_d}{n}\sqrt{1-\left(\frac{v_d}{2\,n\,(V_{GS1}-V_T)}\right)^2}\,\right] \tag{2.10}$$

where K_C is given by

$$K_C = 1 + \frac{1}{1+\dfrac{\beta_4}{\beta_3}} \tag{2.11}$$

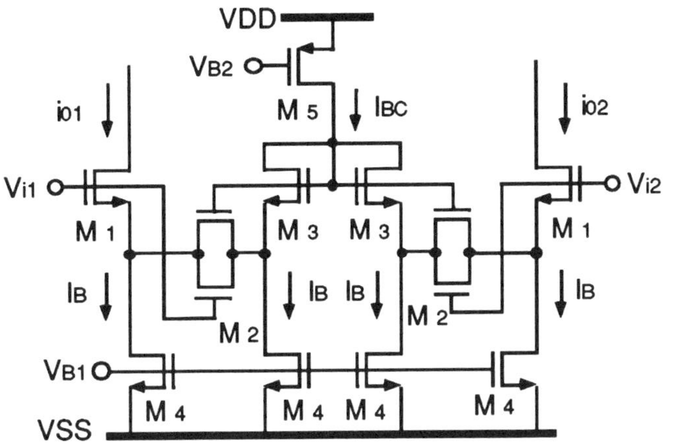

Fig. 2.3. Splitting technique applied to the topology of Fig 2a.

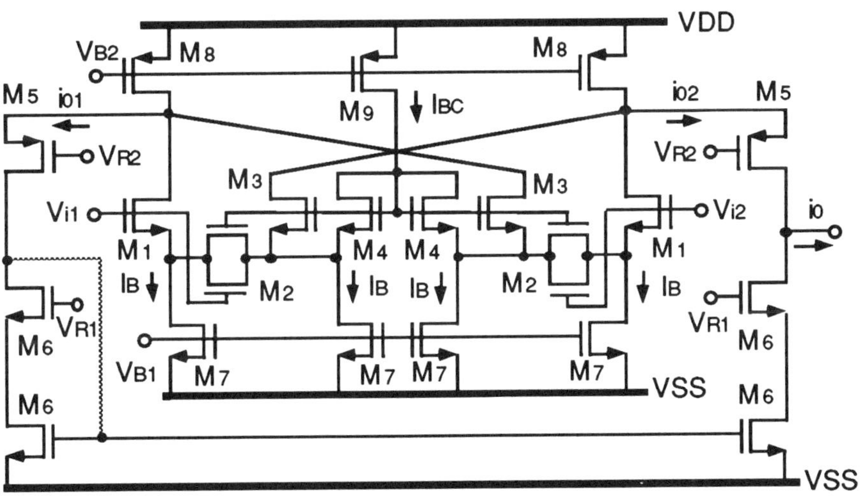

Fig. 2.4. Folded cascode operational transconductance amplifier using the very-low distortion large-signal transconductor.

It is very important to note that factor K_C increases the transconductance but neither V_{GS1} nor THD [24, 25]. Thus, because the transconductance increases by factor K_C and in order to implement the same transconductance as before the factor n can be increased by the same factor. Hence the THD is reduced by a factor $K_C{}^2$. For low distortion applications, expression 2.10 suggests to increase K_C as high as possible. However, this means that β_4 should be reduced. The result is that the feedback decreases. Hence the structure becomes more sensitive to transistor mismatches and this will result in an increment of the even-harmonic distortions. In addition, due to the transistor mismatches the CMRR decreases. Several simulations have shown that the condition $\beta_3=\beta_4$ is a good trade-off, leaving factor K_C equal to 1.5. Under this condition the gate of M_3 and M_4 is the common-mode voltage of the input signal. This voltage has been exploited in fully differential structures [30].

In this topology, the payments for the linearity improvement are lower CMRR, higher noise level, slightly more complex layout and a few percent more silicon area. Furthermore, the bias current increases but the linear range increases as well.

2.3. Second-order effects and frequency response of the Low-Distortion OTA (LDOTA).

In the previous section, the derivation of the transconductor output current, equation 2.10, was based on the ideal quadratic voltage-current relationship. This is, of course, only a first-order approximation. In practice several second-order effects like bulk effects, mobility degradation, finite output conductance and transistor mismatches limit the validity of this equation. Some of these second-order effects are discussed in this section.

2.3.1. Bulk effects. Since the N-well processes used, the bulk of the N transistors is connected to the V_{SS} rail. Even in the case of P-well processes all the N transistors should be allocated in the same well, otherwise the large capacitor from the well to the substrate reduces further the frequency response of the OTA. Due to the large signal, a very important second-order effect is due to the body effect. This body effect affects the V_T of the transistors and the conductance of the triode biased transistors M_2 [2, 8-10]. In order to take into account these effects the transistors M_2 are modeled by the following equation

$$i_{d2} = \left(v_{gs2} - V_{T2} - \left(\frac{1+\eta}{2}\right) v_{ds2} \right) \beta_2 v_{ds2} \tag{2.12}$$

with η and V_{T2} the conductance-degradation factor and the threshold voltage of M_2, respectively. In general, the factor η and V_T are a function of the source-bulk voltage and of the body effect coefficient γ in the form

$$\eta = \frac{d\gamma}{2\sqrt{\phi_B + V_{SBO}}} \tag{2.13}$$

$$V_T = V_{TO} + \gamma\left(\sqrt{\phi_B + v_{SB}} - \sqrt{\phi_B}\right) \tag{2.14}$$

where d is a fitting parameter with value between 0.8 to 0.5 [9]. V_{SBO} is the source-bulk bias voltage and v_{SB} is the source to bulk voltage including ac-signal. For the next analysis the output conductance's of the saturated transistors and the higher order terms are neglected. The output current of the OTA shown in Fig. 2.4, and considering $\beta_3=\beta_4$, can be obtained as

$$i_0 = \Delta I_s + A_1 v_d + A_2 v_d^2 + A_3 v_d^3 + \dots \tag{2.15}$$

with ΔI_s the small dc current due to transistor mismatches. The coefficients A_i are defined in the appendix A2.1. The coefficient A_1 corresponds to the OTA small-signal transconductance and is given by

$$g_m \cong \frac{0.75\sqrt{2\beta_1 I_B}}{1 + \dfrac{\beta_1}{4\beta_2}}\left(1 - \Delta_{T1}\right) \tag{2.16}$$

The term Δ_{T1} reduces the transconductance about 10%. This reduction depends of the technology and of the source-bulk bias voltage (see appendix A2.1 for its definition). As it was discussed in the previous sections, the second and third-harmonic distortions are determined by coefficients A_2 and A_3, respectively. The second harmonic distortion is canceled, except for mismatches, by the current subtraction. HD2 can be easily obtained from the expressions given in appendix A2.1. From relationship 2.3 and appendix A2.1, HD3 becomes

$$\mathrm{HD3} \cong \left[\frac{1 + 12\left(1 + \dfrac{\beta_1}{4\,\beta_2}\right)\left(\eta - \Delta_{T1}\right)}{128\left(1 + \dfrac{\beta_1}{4\,\beta_2}\right)^2 \left(V_{GS1} - V_{T1}\right)^2} + \frac{\dfrac{A_2\,\Delta_{T2}}{A_1} - \Delta_{T3}}{16} \right] V_d^{\,2} \qquad (2.17)$$

It should be noted that for large β_1/β_2 ratios, HD3 still decreases but less due to the effect of both η and Δ_{Ti}. Even more, increasing further β_1/β_2 reduces the OTA transconductance and for the compensation of this reduction it is necessary to increase both bias current and silicon area. This results in a larger transistor output conductance's and larger parasitic capacitors. In this text, values ranging from β_1/β_2=7.5 up to 10 are used.

According to 2.17, using the parameters of the 3 μm N-Well process defined in appendix A1.1 and for $V_d = V_{GS1} - V_{T1}$, HD3 is in the order of -65 dB. HD3 should be larger for P-Well technologies, due to the larger γ factor of the transistors. This is also the case of the scaled down technologies due to the higher concentrations used. Nevertheless, the harmonic distortion is slightly reduced, around 3 dB, due to the internal finite transconductance of transistors M_2. For $V_d = V_{GS1} - V_{T1}$ experimental results for a design using a 1.2 μm process have shown that HD3 is lower than -65 dB, while 2.17 predicts HD3 in the order of -62 dB.

For first-order hand calculations, the bulk effects can be ignored. In this case, 2.17 reduces to

$$\mathrm{HD3} \cong \frac{1}{128\left(1 + \dfrac{\beta_1}{4\,\beta_2}\right)^2 \left(V_{GS1} - V_{T1}\right)^2} V_d^{\,2} \qquad (2.17\ \mathrm{b})$$

This reduced expression is widely used through the whole text.

Other second-order effects. In the LDOTA, there are other second-order effects like transistor mismatches and mobility degradation. The transistor mismatches unbalance the topology, avoiding the exact cancellation of the even-order distortions.

In a single-ended OTA, Fig. 2.4, simulated results for input voltages of 2.4 Vpp with 2% transistor mismatches have shown HD2=0.15% (-56.5 dB) and HD3 = 0.11% (-59.2 dB). With 1% transistor mismatches HD2 is reduced to -62 dB and HD3 to -60.6 dB. In current CMOS processes, transistor mismatches of 1% are common. The experimental results are in good agreement with the assumption of 1% transistor mismatches. Hence, the effects of transistor mismatches are small, showing that the LDOTA is little sensitive to these effects. The mobility degradation acts in the transistors like additional source degeneration. The transconductance decreases but the linearity is not further degraded [23]. For the transistors involved in the voltage-to-current transducer, a minimum channel length of 6 μm is very often used. It reduces the mobility effects and also reduces the output conductance of the transistors.

2.3.2. Differential-mode Frequency response. In this section the frequency response of the low-distortion OTA input stage is considered. In the OTA of Fig. 2.4, the effect of the parasitic poles due to the cascode transistors and those due to the differential-to-single ended converter are not taken into account. For the differential-mode transconductance only the dominant zero and the dominant pole are considered. Furthermore, the gate capacitor and the effect of the transconductance of M_2 are neglected. If finally the conductance in the common node, gates of M_3 and M_4, is assumed to be small the differential-mode transconductance g_{dm} is approximately given by

$$g_{md} \cong g_m \frac{1 + \dfrac{s}{\omega_{zd}}}{1 + \dfrac{s}{\omega_{pd}}} \tag{2.18}$$

with

$$g_m \cong \frac{3\, g_{m1}\, g_{o2}}{g_{m1} + g_{mb1} + 4\, g_{o2}} \tag{2.19}$$

$$\omega_{zd} \cong \frac{3\, g_{o2}\left(g_{m1} + g_{mb1}\right)}{C_2\left(g_{m1} + g_{mb1} + 2g_{o2}\right) + \left(C_3 + C_5\right)\left(2g_{o2}\right)} \tag{2.20}$$

$$\omega_{pd} \cong \frac{(g_{m1} + g_{mb1})(g_{m1} + g_{mb1} + 4\,g_{o2})}{(g_{m1} + g_{mb1} + 2g_{o2})(C_1 + C_2 + C_3 + C_5)} \tag{2.21}$$

where g_m is the small-signal transconductance of the voltage-to-current transducer. g_{mbi} is the bulk transconductance of the ith transistor and g_{o2} is the drain-source conductance of the triode biased transistors M_2. The transconductance g_m is equivalent to the transconductance already defined in expression 2.16. The transistor transconductances are such that $g_{m1}=g_{m3}/2=g_{m4}/2$. For the bulk transconductance it is considered $g_{mb1}=g_{mb3}/2=g_{mb4}/2$. The capacitance C_1 is the gate-source capacitor of M_1 and C_5, almost of the same value as C_1, is the sum of the gate-source capacitor of M_3 and M_4. C_2 and C_3 are the total parasitic capacitors from the source of M_1 and M_3 to ac grounds, respectively. For typical low distortion designs g_{o2} is much smaller than g_m. Under these conditions, the zero and the pole of the differential frequency response can be approximated by

$$\omega_{zd} \approx \frac{g_m}{C_2} \tag{2.22}$$

and

$$\omega_{pd} \approx \frac{1.33\left(1 + \dfrac{\beta_1}{4\,\beta_2}\right)}{C_1 + C_2 + C_3 + C_5}\,g_m \tag{2.23}$$

It should be remarked that these equations have been obtained for very simple hand calculations. Due to the approximations, the error in the computations can be as high as 30%. Nevertheless, these expressions allow us to visualize the frequencies of the pole-zero pair. Typically, the frequency of the pole is higher than that of the zero. Thus, this pole-zero pair introduces excess lead phase in the small-signal transconductance. This is not necessarily true for the whole OTA wherein other parasitic poles must be considered. From 2.22 and 2.23, it is clear that for high-frequency applications it is mandatory to reduce the parasitic capacitors of the low-distortion input stage (C_1, C_2, C_3, C_4 and C_5).

2.3.3. Common-mode frequency response. Typical ac analysis for the common-mode equivalent circuit leads to the following single-side common-mode transconductance

$$g_{mc} \cong \frac{g_{m1}\, g_2 + g_{mb1}\left(g_3 + g_4\right)}{g_{m1} + g_{mb1} + g_2} \frac{1 + \dfrac{s}{\omega_{zc}}}{1 + \dfrac{s}{\omega_{pc}}} \tag{2.24}$$

with

$$\omega_{zc} = \frac{g_{m1}\, g_2 + g_{mb1}\left(g_3 + g_4\right)}{\left(C_2 + C_3\right) g_{m1} + C_4\left(g_{m1} + g_{mb1}\right)} \tag{2.25}$$

$$\omega_{pc} = \frac{\left(g_{m1} + g_{mb1}\right) g_{m1}\, g_{o2}}{\left(C_1 + C_2\right) g_{m1}\, g_{o2} + \left(\dfrac{C_3\, g_{m1}}{2} + C_4\, g_{m1} + \dfrac{C_5\, g_{mb1}}{2}\right)\left(g_{m1} + g_{mb1} + 4\, g_{o2}\right)}$$

(2.26)

where g_{mc}, ω_{zc} and ω_{pc} are the common-mode transconductance, the frequency of the zero and the frequency of the pole, respectively. The conductance g_2 and g_3 are the total grounded conductance connected to the source of M_1 and M_3, respectively. The capacitor C_4 is half of the total parasitic capacitance between the gate of M_4 and ac ground. For low distortion design wherein $g_{m1} >> g_{o2}$ and neglecting g_{mb} , more simple but less accurate relations can be deducted. Thus, the low-frequency common-mode transconductance is dominated by g_2. This conductance is basically the output conductance of the transistor M_7, see Fig. 2.4. The dominant zero and dominant pole become

$$\omega_{zc} \approx \frac{g_2}{C_2 + C_3 + C_4} \tag{2.27}$$

$$\omega_{pc} \approx \frac{1}{\dfrac{C_1 + C_2}{g_{m1}} + \dfrac{\dfrac{C_3}{2} + C_4}{g_{02}}} \tag{2.28}$$

The frequency of the zero is dominated by g_2 and typically is lower than the differential unity gain frequency of the integrators by a factor 10 or even less. It should be said that this effect is common in all the circuits using a differential pair as input stage. The frequency of the common-mode pole is of the same order of magnitude as the frequency of the differential-mode pole. From equations 2.18 to 2.28, the low-frequency single-side Common-Mode Rejection Ratio (CMRR) of the voltage-to-current transducer, the ratio of g_{md} to g_{mc} for a single-side current is

$$\text{CMRR (single side)} \approx \left(\frac{0.75}{1 + \dfrac{\beta_1}{4\,\beta_2}} \right) \left(\frac{g_{m1}}{2\,g_2} \right) \frac{1 + \dfrac{s}{\omega_{zd}}}{1 + \dfrac{s}{\omega_{zc}}} \qquad (2.29)$$

In the single differential pair, the dc single-side CMRR is given by g_m/g_2. In the case of the very linear transconductor, this is reduced by a factor ≈ 2. At medium frequencies, the behavior is dominated by the pole at frequency ω_{zc}. Then the CMRR reduces by 20 dB per decade until the frequency where the effect of the zero compensates the effect of the pole. It can be concluded that the low-frequency single-side CMRR is mainly limited by g_2, as can be expected. Simulated and calculated results for a 150 μA/V transconductor have shown a dc CMRR in the order of 20 dB for a single-side output current and a pole frequency ω_{zc} in the order of 200 kHz. Furthermore, this type of transconductors are used in a full OTA wherein current or voltage subtraction takes place at the output. In the ideal case, the common-mode signals are present at both transconductor outputs. Hence, after the subtraction, these signals are ideally canceled. In the real case, due to mismatches the common-mode signals are not fully canceled but they are reduced in the order of -30 to -40 dB.

2.4. Rejection to supply noise and noise level.

The importance of the reduction of the noise coupled from the supply rails has been pointed out by several authors [26 - 28]. Due to the finite impedance of the supply sources, non-zero resistance of the metal lines in the chip and the coupling capacitors noise is injected to the OTA through the V_{DD} and V_{SS} rails. The rejection of the OTA to this noise is measured by the Power-Supply Rejection Ratio (PSRR) parameter. PSRR is defined as the ratio of the small-signal differential-mode transconductance to the small-signal differential-mode transconductance from the supply rail. The last transconductance is defined as the ratio of the OTA output current to the supply rail applied ac-signal. There are two PSRR in the OTA, one for each supply rail. An analysis of these parameters is carried out in the next two sections.

2.4.1 Rejection to V_{DD} noise. In the LDOTA discussed in the previous sections (see Fig. 2.4) the V_{DD} noise is coupled to the OTA output through the transistors M_5, M_8 and M_9. A simplified circuit showing these coupling paths is depicted in Fig. 2.5. It is assumed that the gate voltages of the PMOS transistors are generated from the positive supply source. Thus, they are ac tied to V_{DD}. In order to solve the circuit for the output current, we can take advantage of the symmetry of the circuit. So, it is sufficient to analyze one half of the circuit. Lets consider first the current due to transistors M_5 and M_8. In Fig. 2.6a, a small-signal single-side equivalent circuit is shown. In this figure, M_5 is a capacitor free transistor. The equivalent capacitors and conductance's are defined as follows

$$C_8 = C_{DG8} + C_{DB8} + C_{GS5} + C_{SB5}$$
$$C_I = C_{DG1} + C_{DB1} + C_{DG3} + C_{DB3}$$
$$C_5 \cong C_{DG5} + C_{DB5}$$
$$g_I \cong g_{DS1} + g_{DS3}$$

In the last expression, the conductance g_I represents the grounded conductance due to the input stage. In Fig. 2.4, this impedance is due to the finite output conductance of M_1 and M_3. Because these conductance's are connected to the source of their own transistor, for low frequencies there is a cascode effect. This effect is not include in g_I, otherwise the equations become rather involved. The equivalent circuit shown in Fig. 2.6a can be represented by a Norton equivalent circuit as shown in Fig. 2.6b. The computation of the equivalent current source leads to

$$\frac{i_{eq}}{v_{dd}} = \left(\frac{g_5\, g_8 - g_{m5}\, g_I}{g_{m5}}\right) \cdot \left(\frac{1 + s\dfrac{g_5\, C_8 + (g_I + g_5 + g_8) C_5 + g_{m5}(C_5 - C_I)}{g_5\, g_8 - g_{m5}\, g_I} + s^2 \dfrac{C_5 (C_I + C_8)}{g_5\, g_8 - g_{m5}\, g_I}}{1 + s\dfrac{C_I + C_8}{g_I + g_5 + g_{m5} + g_8}} \right) \tag{2.30}$$

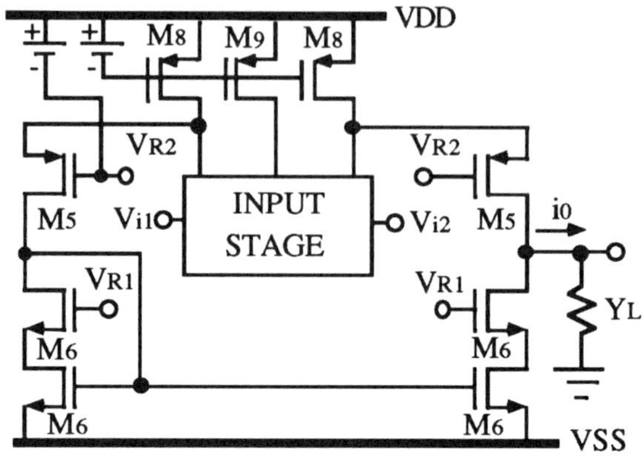

Fig. 2.5. Simplified OTA diagram for the v_{dd} transconductance.

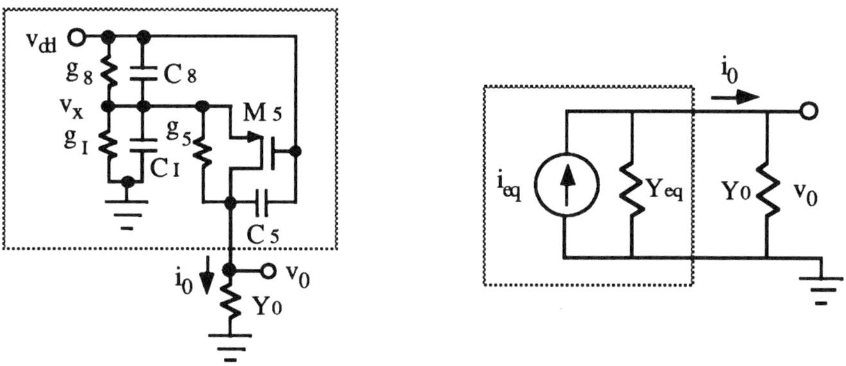

Fig. 2.6. Small-signal circuit for a single-side v_{dd} transconductance.
a) equivalent circuit and b) Norton representation.

The dc value of expression 2.30 is given by the term $(g_5\, g_8 / g_{m5}) - g_I$. The dominant zero is dominated by the terms $g_5\, C_8 + g_{m5}\, C_5$. This is because the capacitor C_8 includes larger capacitors than C_5. Typically, C_8 is 3 - 5 times larger than C_5. However, for high-frequency applications, a huge current can flow through M_8 which increases g_8. Then the term $C_5\, g_8$ can not always be neglected. The non-dominant zero is placed at very high frequencies, typically in the range of hundred MHz. The frequency of this zero is slightly larger than the frequency of the dominant pole. For hand calculations, 2.30 can be approximated into

$$\frac{i_{eq}}{v_{dd}} \cong \frac{g_5\, g_8 - g_{m5}\, g_I}{g_{m5}} + s\left(\frac{g_5\, C_8 + (g_5 + g_8)\, C_5}{g_{m5}} + C_5 - C_I\right) \tag{2.30b}$$

From Fig. 2.6a, the Norton admittance can be approximately obtained as

$$Y_{eq} \cong g_{eq} + s\, C_{eq} = \frac{g_5\, g_8}{g_{m5}} + s\left(\frac{g_5\,(C_I + C_5 + C_8) + g_8\, C_5}{g_{m5}} + C_5\right) \tag{2.31}$$

According to this expression, the low-frequency behavior of Y_{eq} is dominated by the conductance of the cascode transistors M_5 and M_6, $g_5\, g_8 / g_{m5}$. The high-frequency behavior is dominated by the capacitor C_5. This equivalent circuit allows to calculate the PSRR of the OTA. Using the equivalent Norton circuit to each side of the OTA, the circuit becomes as shown in Fig. 2.7.

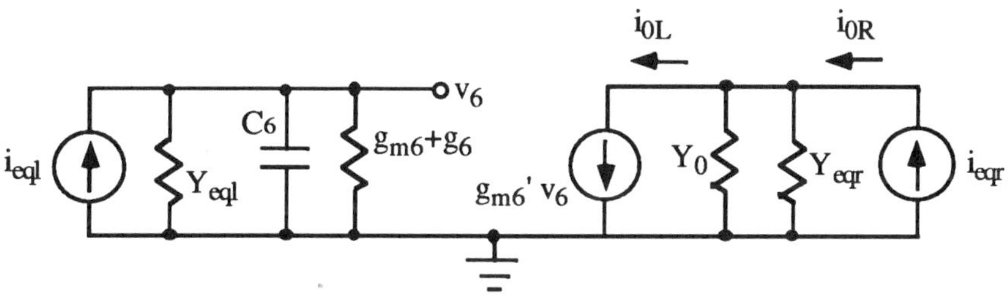

Fig. 2.7. OTA equivalent circuit for the v_{dd} transconductance.

For the right-hand side circuit, the current, i_{eqr} in Fig. 2.7, is injected directly to the OTA output. For the left-hand side, i_{eql} flows through a current divider circuit. This current divider is due to the gate-drain connection of the transistors M_6 and their grounded impedances associated. Afterwards, the drain current of the left-hand side bottom transistor M_6 is mirrored to the OTA output, see Figs. 2.5 and 2.7. Detailed computations show that the total output current is approximately given by

$$\frac{i_0}{v_{dd}} = \frac{i_{0R} - i_{0L}}{v_{dd}} \cong$$

$$\cong \left(\frac{g_5\, g_8}{g_{m5}} - g_I\right) \left(\frac{g_{eq}\, g_6}{g_{m6}} + \Delta_s\right) \left(1 + s\frac{C_5}{\dfrac{g_5\, g_8}{g_{m5}} - g_I}\right) \left(\frac{1 + s\dfrac{C_{eq} + C_6}{g_{eq} + \Delta_s\, g_{m6}}}{1 + s\dfrac{C_{eq} + C_6}{g_{m6}}}\right) \qquad (2.32)$$

where $g_{eq} \cong g_5\,(g_I + g_8) / g_{m5}$. The factor Δ_s is a result of the mismatches in both the current mirror and the equivalent current sources at both sides of the OTA and is given by

$$\Delta_s = 1 - \frac{(g_{m6})_{right}}{(g_{m6})_{left}} \frac{i_{eql}}{i_{eqr}} \qquad (2.32b)$$

The mismatches in transistors M_6 can be as small as 1 %. Nevertheless, the mismatch between the left-hand side current i_{eql} and the right-hand side equivalent currents is much higher, especially at high frequencies. This is due to the effect of the voltage-dependent drain-bulk capacitor of M_5, see equation 2.30. This capacitor mismatch is function of the gamma factor and also a function of the drain-bulk bias voltage. Due to this effect the mismatches on Δ_s are in the order of 4 - 8 % at medium and higher frequencies.

In equation 2.32, four different regions can be distinguished for the v_{dd} transconductance. At low frequencies, the behavior is dominated by the factor Δs ($g_5\, g_8 - g_I\, g_{m5}$) / g_{m5}. At frequencies higher than the first zero but lower than the second zero, the behavior of the v_{dd} transconductance is capacitive. In this region, the v_{dd} transconductance can be approximated by a capacitive admittance equal to $s\,\Delta_s\,C_5$. It can be noted that this behavior is a result of the OTA mismatches. At frequencies higher than the second zero but lower than the high-frequency pole the v_{dd} transconductance can be approximated by $s^2\, C_5(C_5 + C_6)/g_{m6}$. This effect is due to the excess phase introduced because the additional pole at the left-hand side of the circuit. At this frequency range, the v_{dd} transconductance increases by 40 dB by decade. In this region, the v_{dd} transconductance is not affected by the mismatches but by the absolute value of the capacitors C_5 and C_6 and the transconductance of M_6. Finally, at very high frequencies, the v_{dd} transconductance is dominated by the capacitor C_5. In this region, the left-hand side current is much smaller than the right-hand side current. Hence, most of the output current is due the right-hand side capacitor C_5.

A second current component is due to both output conductance and drain capacitors of M_9, see Figs. 2.4 and 2.5. Assuming that the impedance at the gate of transistors M_3 and M_4 is very small compared with g_9 and C_9, this current can be approximated by

$$\frac{i_0}{v_{dd}} \cong \frac{\Delta g_9 + s\,\Delta C_9}{2} \tag{2.33}$$

Typically this current is much smaller than the current of expression 2.32. These relationships, 2.32 and 2.33, are in good agreement with the Spice results.

2.4.2. Rejection to V_{SS} noise. The computation of the output current due to the v_{ss} noise is rather involved because the several signal coupling paths. A part of the output current is generated because the 4 current sources biasing the OTA input stage. The other current component is due to transistors M_6, especially those transistors at the OTA output stage. A detailed circuit analysis of the OTA input stage shows that the single-side current generated by a signal applied to V_{SS} can be approximated by a high-pass transfer function with a transconductance given by

$$\frac{i_{01}}{v_{ss}} \cong g_7\left(1+\frac{\beta_1}{4\,\beta_2}\right)\left(1+s\frac{C_7}{g_7}\right) \tag{2.34}$$

where g_7 and C_7 are the output conductance and the drain capacitors of M_7, respectively. From this expression, it is clear that the high-frequency limitation is not the output conductance of the current sources but the drain capacitors of M_7, drain-bulk and drain-gate. These capacitors can be reduced by reducing the drain dimensions of transistors M_7. If the LDOTA is compared with a differential pair input stage OTA, the single-side LDOTA v_{ss} transconductance due to the input stage increases by a factor n, the linearization factor. This is the price to pay for the lower distortion in the voltage-to-current transducer. However, it should be noted that this current is almost purely common-mode signal and so it is further reduced at the output of the OTA after the subtraction. Assuming that C_7=1 pF, β_1/β_2 = 10, and g_m=150 μA/V, the single-side current gives a single-side PSRR of the order of 13 dB at 1.5 MHz. If 1 % transistor mismatches are included, at the output of the OTA the current should be reduced by 40 dB leading to a PSRR ≈ 53 dB. It has to be remembered that this value is only due to the input stage transistors.

The second current component is due to the transistors M_5 and M_6. Typically, these are the most important coupling paths for the v_{ss} noise. This coupling is present in all cascode topologies [28]. The first path is due to the transistors M_5 and M_6 at the left-hand side of the circuit, see Fig. 2.4. The transistors M_6 are connected in a feedback loop leading to a resistor in parallel with a capacitor. The grounded impedance at the gate of the bottom transistor M_6 is approximately given by $(g_5 g_8)/g_{m5} + s(C_{dg5} + C_{db5})$. This grounded impedance produces drain current on transistors M_6 and so it appears at the output of the OTA. The second coupling is due to the drain - gate and drain - bulk capacitors of the M_6 right-hand side transistors. The situation is very like this analyzed in Fig. 2.5. It can be shown that the drain current is generated by a conductance given by $g_6^2/g_{m6} + s\,(C_{dg6} + C_{db6})$. Adding both current components, the v_{ss} transconductance becomes

$$\frac{i_0}{v_{ss}} \cong \frac{g_5\, g_8}{g_{m5}} + \frac{g_6^{\,2}}{g_{m6}} + s\left(C_5 + C_6\right) \tag{2.35}$$

with the capacitor $C_X = C_{DGX} + C_{DBX}$. In reality, there is a very low-frequency pole and zero which almost cancel each other and so, this effect can be neglected. The total output current generated by v_{ss} can be characterized by a total v_{ss} transconductance given by

$$g_{mvss} \cong \frac{g_5\, g_8}{g_{m5}} + \frac{g_6^{\,2}}{g_{m6}} + \Delta\, g_7\left(1 + \frac{\beta_1}{4\,\beta_2}\right) + s\left(C_5 + C_6 + \Delta\, C_7\left(1 + \frac{\beta_1}{4\,\beta_2}\right)\right) \tag{2.36}$$

From this expression it can be seen that the dc transconductance is mainly determined by the output conductance of the current sources used in the biasing of the input stage, g_7. Its effect is further reduced by the factor Δ (≈ 0.02 - 0.01). Also, at low frequencies, the v_{ss} transconductance is affected by the finite output conductance of the transistors M_5 and M_6. At medium frequencies, the behavior is dominated by the capacitors C_5 and C_6. For the design values used in the previous section and considering $C_5 + C_6 = 0.5$ pF, a v_{ss} transconductance of 4.5 μA/V is obtained at 1.5 MHz. For the 150 μA/V transconductor, the v_{ss} transconductance represents a PSRR = 30.5 dB. This result is in good agreement with the PSRR = 30 dB experimentally obtained.

In general, if larger PSRR is needed, it is mandatory to reduce the drain capacitors of the transistors M_5 and M_6. This can be done by reducing as much as possible both their dimensions and their drain area. However, following this approach, the improvement on the PSRR is limited. A better solution is to use fully-differential structures, to be discussed in the next section.

2.4.3. Noise level of the low-distortion OTA. In general, the range of signals that can be accurately driven by electronic devices is limited. For low signals, the minimum amplitude that can be processed is limited by the electrical noise. Noise is considered as all the unwanted electrical signals generated in the device or externally and coupled to the output of the system. These signals are present in the system whether input signals are applied or not. The noise signals interfere with the incoming signal and make impossible to detect with sufficiently quality signals with an amplitude comparable to the noise level. Moreover, signals below this level are almost impossible to be detected. So, noise in the device or in the system represents the lowest level for the incoming signal.

The origins of noise can be classified as intrinsic transistor noise sources and extrinsic noise sources. In CMOS transistors, the intrinsic noise is generated by different mechanisms. These mechanisms are not further discussed here but the reader can find excellent descriptions in the references [8, 29]. There are two dominant noise sources in CMOS transistors, namely, channel thermal noise and 1/f or flicker noise. The extrinsic noise is mainly due to the signals produced by the surrounding circuitry and coupled to the device or to the system. The reduction of the effects of the extrinsic noise has been considered in the previous sections. In this section, the noise generated by the transistor is considered.

Thermal noise. The flow of carriers across the channel is due to the drain-source electric field and because the inversion area under the gate. The channel, due to the gate-source voltage, acts like a resistor. Thus, due to the random flow of the carriers thermal noise is produced. It can be demonstrated that the spectral density of the thermal noise drain current in a CMOS transistor biased in the linear region, $V_{DS} < V_{GS}-V_T$, can be approximated by [29]

$$i_d^2 = \frac{8}{3} k\, T \mu\, C_{ox} \frac{W}{L} \left(\frac{3 (V_{GS} - V_T) V_{DS} - 3 (V_{GS} - V_T)^2 - V_{DS}^2}{2 (V_{GS} - V_T) - V_{DS}} \right) \qquad (2.37)$$

where k, T, μ and C_{ox} are the Boltzmann constant, the temperature (in Kelvin degrees), the channel mobility and the gate capacitance per unit of gate area, respectively. W and L are the width and the length of the transistor gate. Typically, when the transistor is biased in linear region the voltage V_{DS} is very small, then 2.37 becomes

$$i_d^2 = 4\, k\, T \left(\mu\, C_{ox} \frac{W}{L} (V_{GS} - V_T) \right) \cong 4\, k\, T\, g_0 \qquad (2.37\ b)$$

with g_0 the small-signal conductance of the transistor. This spectral noise density corresponds to that of a resistor with a resistance value equal to $1/g_0$. In saturation region, $V_{DS} > V_{GS}-V_T$, the noise can be calculated from 2.37b but with g_0 equal to $2g_m/3$, with g_m the small-signal transconductance of the transistor.

1/f noise. The origins of this type of noise are not yet fully understood. Most of the theories explain this kind of noise in function of the imperfections in the silicon-silicon oxide interface. The surface states and the traps in this interface randomly interfere with the charges flowing through the channel. Due to this fact, the generated noise is strongly dependent of the technology. Furthermore, it is inversely proportional to the gate dimensions because in larger gate areas more traps and surface states are present and certain averaging occurs. Because the technology dependence of this type of noise, this is commonly expressed as function of a constant K_F which is a fitting parameter. So, the spectral density of the 1/f noise drain current is commonly characterized in the form [29]

$$i_d^2 = \frac{K_F I_{DS}}{C_{ox} L^2 f} \qquad (2.38)$$

where I_{DS} is the bias drain current and f is the frequency, in Hertz. The spectral noise density can be expressed as an input referred spectral noise voltage. This is a more useful relation for the circuit designer because the signal resolution is commonly measured at the input of the systems. The input referred spectral density of the noise of a transistor biased in linear region can be easily calculated as

$$v_g^2 = \frac{i_d^2}{g_m^2} = \frac{KF}{2\mu C_{ox}^2 WLf} \tag{2.39}$$

with i_d^2 characterized by expression 2.38. This relationship will be widely used in this text.

Noise of the low-distortion voltage-to-current transducer. For the low-distortion transducer, the output referred noise current can be calculated for each transistor of the transconductor by well known techniques [8,10,29]. This calculation is tedious and so only the results are presented. Table 2.1 shows the approximated contribution of the transistors to the output squared noise current. In the first column, the number in brackets is the number of transistors taken in to account in the calculation. The noise contributions are normalized by the output noise current generated by a single differential pair transconductor, for the same transconductance. This ratio is called the Noise Factor, NF. The values of the second column are the Squared Noise Factor, NF^2, of every group of transistors. The square root of the sum of these squared factors gives the noise factor of the low-distortion transducer. For the computed squared noise factors, the beta of the transistors are such that $\beta_1 = \beta_7 = 10\ \beta_2$.

From table 2.1, it is clear that the major noise contributions are those due to transistors M_2, M_3 and M_7. For low-distortion applications, β_1 is in the order of $8\beta_2$ to $10\beta_2$, the squared noise factor of the transistors M_3 and transistors M_7 are in the order of 2.4 and $0.6(\beta_7/\beta_2)$, respectively. Typically β_7/β_2 is higher than 10. Hence, the noise is dominated by the transistors M_7, instead of the transistors involved in the voltage-to-current conversion. This is a very important design consideration: the larger the transconductance of M_7 the smaller $V_{GS7} - V_T$ and V_{DSAT7} are. Hence, a larger OTA common-mode range is obtained. However, the larger the transconductance of M_7 the larger the noise contribution of M_7, $i_d^2 = (8/3)kTg_{m7}$.

transistors	squared noise factor (NF^2)	NF^2
(2) M_1	$\frac{6}{4}\left(\frac{1}{1+\frac{\beta_1}{4\beta_2}}\right)$	0.43
(4) M_2	$\frac{9}{2}\left(\frac{1}{1+\frac{4\beta_2}{\beta_1}}\right)$	3.2
(2) M_3	$\frac{1}{48}\left(\frac{\left(10+\frac{\beta_1}{\beta_2}\right)^2}{1+\frac{\beta_1}{4\beta_2}}\right)$	2.4
(2) M_4	$\frac{1}{48}\left(\frac{\left(-2+\frac{\beta_1}{\beta_2}\right)^2}{1+\frac{\beta_1}{4\beta_2}}\right)$	0.38
(4) M_7	$\frac{8}{27}\left(\frac{\beta_1\beta_7}{\beta_2^2}\right)\left(\frac{1+\frac{\beta_1}{4\beta_2}}{\left(1+\frac{\beta_1}{3\beta_2}\right)^2}\right)$	$\cong 6$

Table 2.1. Normalized spectral density of the noise output current for the transistors of the low-distortion transducer. The contribution of each set of transistors has been normalized by the spectral density of the noise output current of a single differential pair. For the NF^2 values the conditions $\beta_1=\beta_7=10\beta_2$ have been used.

For transistors M_7, there is a trade off between noise and OTA common-mode range. The physical reason of the large noise contribution of these transistors can be explained as follows: due to the source degeneration involved in the low-distortion voltage-to-current transducer, the noise drain current of the left most and right most transistors M_7 are directly injected to the output of the OTA through only one branch, see Fig. 2.4. This current is almost differential, so this noise current is not reduced at the output. This is not the case for both transistors M_7 connected to transistors M_3 and M_4, see Fig. 2.4. In this case, the noise drain current is split in both branches of the OTA input stage. Hence, this noise contribution is reduced at the output of the OTA. These considerations are further used in the next chapters for the design of high-performance filters.

2.5. Design considerations for Fully-Differential OTAs.

In the previous section, a single-ended very low distortion OTA has been discussed. Hand calculations have shown that is feasible to design OTAs with THD lower than -60 dB for an input signal larger than 2.4 Vpp. Due to the unavoidable transistor mismatches and its asymmetries a second-order distortion in the order of -62 dB is commonly found. In the next chapter experimental and simulated results for a single-ended OTA are presented. Those results show HD3 in the order of -65 dB for input signals up to 2.4 Vpp. However, the THD is dominated by HD2. In addition, due to the asymmetries of the topology the rejection to common-mode signals is in the order of 30 to 40 dB.

Several techniques for increasing the rejection to common-mode signals and supply signals have been reported [26-27]. Some of these techniques minimize the coupling capacitors between the substrate and the signal paths. The major disadvantage of this technique is that the transistors of the OTA input stage have to be fabricated in a different well with their sources connected to the well. So, this technique can not be used in N-Well technologies for input stages using NMOS transistors. Furthermore, this technique increases the parasitic capacitors, due to the well at the source terminal of the transistors. The result is a degradation of the CMRR at high frequencies. Even more, the major limitation of the negative PSRR, as previously discussed, is due to the capacitors associated with the OTA output stage. For higher rejection to supply noise, the most commonly used technique uses differential signal paths instead of common-mode voltage referred signal paths. This section deals with the design of this type of topologies. The advantages and drawbacks of the fully-differential structures are discussed in section 2.5.1. The major design problems of these topologies are addressed in section 2.5.2 . A quite efficient technique for the design of low distortion fully-differential OTAs is described in section 2.5.3.

2.5.1. Fully-Differential topologies. A fully-differential version of the low-distortion OTA discussed in the previous section can be obtained by removing the soft line at the left-hand side of the topology shown in Fig. 2.4. Thus, the left-hand side of the topology is enabled as a second output voltage. This leads to a topology with 2 inputs and 2 outputs. That topology is shown in Fig.2.8. Ideally, in the fully-differential and fully-symmetric topology all the common-mode signals and even-order distortions should be present with the same amplitude and same phase at both output terminals of the structure. Hence, these signals should be canceled each other by simple subtraction, leading to topologies with higher CMRR and higher PSRR. A discussion on these structures follows.

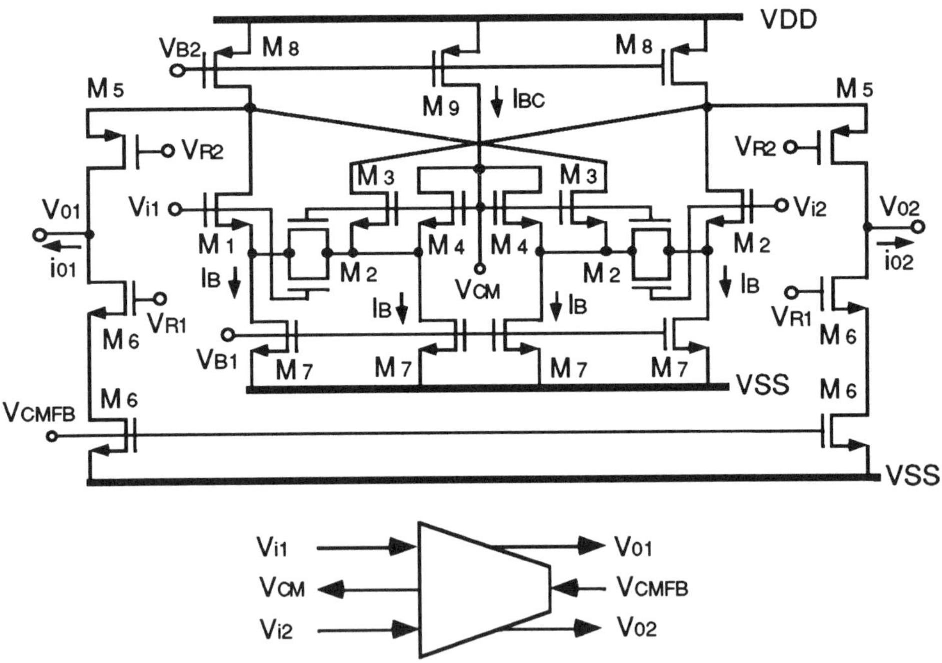

Fig.2.8. Fully-differential low-distortion OTA with common-mode detection.

Dynamic range. The dynamic range is determined by the ratio of the maximum swing of the differential input signal with low distortion to the input referred noise level. Hence, for higher dynamic range these parameters, differential input range and noise level, have to be increased and reduced, respectively. In single-ended OTAs, the signals are always referred to the common-mode voltage. In this case, the ac input signal is applied to one input of the OTA while the other one is commonly connected to the common-mode voltage reference. If a sinusoidal signal is considered, $v_i = V \sin(\omega t)$, for the OTA half of the signal is common-mode signal and only half of the signal is differential. The common-mode swing of the OTA input stage is, however, equal to V. This is not the case for fully-differential signals. In this case, the ac-signal is present at both input terminals of the OTA. Most of the signal is differential and the common-mode signal is quite small. So, for the same common-mode range, the swing of the differential signal is twice that of the single-ended OTA. The dynamic range, proportional to the maximum swing of the input signal, is improved by 6 dB. In the real case, the dynamic range improvement is lower than this value because the noise level of the fully-differential topology is slightly higher than for single-ended topologies. The reason is the additional circuitry used for common-mode stabilization.

Rejection to common-mode signals. The output current of the voltage-to-current transducers is a differential signal. In single-ended topologies, it is converted to a single-ended signal by mirroring and subtracting one half of the transducer output current. Due to this conversion, additional poles and mismatches are introduced in one branch of the OTA. The additional poles introduce excess lag phase in the current flowing in that branch and the additional mismatches in the current mirrors introduce errors in the magnitude of the current. As a result, the common-mode signals are not in phase and with the same amplitude, thus avoiding their exact cancellation. In fully-differential topologies both branches can be designed fully-symmetrical. Therefore higher rejection to common-mode signals can be obtained. Also, another quite important benefit is the further reduction of the even-order non-linearities, which in fact are common-mode signals. As a result, lower THD, higher CMRR and higher rejection to supply noise are obtained.

Common-mode stabilization. The major advantages of fully-differential structures are that the signal is differential instead of referred to a ground. The major disadvantages of these techniques are also associated to this fact. Because the fully-differential nature of the system, the common-mode voltage of the topology (the operation point) can not be stabilized by the differential feedback. Thus, in order to define the common-mode voltage of the structure, additional circuitry has to be included. The loop that determines the operation point of the structure is the Common-Mode FeedBack (CMFB).

The design of the CMFB increases the complexity of the OTA because two signal paths must be considered. The first path is the normal signal path that defines the transfer function of the system: differential input signal to differential output signal. The second path is due to the common-mode signals, determined mainly by the CMFB used for common-mode stabilization. Typically, the design of the common-mode loop is not easy because the output signal is mainly differential and the common-mode signal must be detected as fast as possible with simple and fast circuitry. Furthermore, harmonic distortion generated in this loop introduces additional distortion in the differential loop.

Additional disadvantages. Due to the additional loop for common-mode stabilization, additional circuitry must be included in the OTA. This means, more silicon area, slightly more complex layout and more power consumption. The increment of the silicon area is not a very serious problem, typically in the order of 20%. Nevertheless, depending of the implementation of the common-mode loop, the power consumption could increase between 50 to 100 %. In section 2.5.3, a technique is discussed that reduces drastically the complexity of the design, the silicon area and the power consumption .

2.5.2. Common-mode stabilization. The most important advantages of differential structures are due to the differential nature of the signal. As it has already been mentioned, by taking the output in a differential form most of the common-mode signals are canceled. This does not mean that the common-mode signals at each OTA output are zero. The even-order distortions and the other common-mode signals are present at both outputs, in the ideal case with the same amplitude and same phase. One of the major questions of the designers of fully-differential structures is how to suppress these signals? More important, how can the quiescent point of the OTA be accurately controlled? In the remainder of this chapter, it is tried to answer these questions.

In the fully-differential OTA, the mismatches in the biasing current should unbalance the topology and make that some transistors go out of saturation region. In order to clearly see the effects of the common-mode signals lets consider a simple common-mode excitation at the input of a grounded capacitances loaded OTA. The common-mode signal could be noise coupled to the outputs of the OTA. This signal produces a current at both OTA output nodes and so the differential output signal $v_d=v_{01}-v_{02}$ is zero if the mismatches are neglected. For the common-mode output voltage, a transfer function can be defined as

$$A_{cc} = \frac{v_{0c}}{v_{ic}} \cong \frac{g_{mc}}{g_0} \frac{1}{1 + s\dfrac{C_0}{g_0}} \tag{2.40}$$

with C_0 the OTA output connected grounded capacitor and g_0 the OTA output conductance. The transconductance g_{mc} depends of the input stage being considered but in most of the cases has the form of equation 2.24. This analysis can be applied for the supply noise if the appropriate common-mode transconductance is used. Since at the output of the OTA cascode transistors are used, the OTA output conductance g_0 is very small, e.g. less than 1 μA/V. Thus, the common-mode dc gain could be in the order of 50 to 100. The relationship 2.40 shows a low-pass behavior. It should be remarked that the common-mode transconductance, g_{mc}, has a low-frequency zero. This zero compensates the effect of the pole in 2.40. The result is a transfer function whose high-frequency behavior depends of the parasitic and load capacitors.

The meaning of the high common-mode dc gain is that the operation point of the OTA is very sensitive to the common-mode signals. Hence, any offset current could produce a shift in the operation point of several Volts, and so saturating the OTA. In practice, mismatches in the bias currents can easily produce offset currents in the order of several tenths of μA. It should also be recognized that the even-order harmonics produce common-mode dc signals. As a result, the operation point of the OTA is quite unstable. In addition, the low-frequency common-mode signals are also further amplified. These signals increases the swing of the OTA output signal and produces premature clipping. Further, the swing of the input signal for the next OTA should be larger, increasing further the distortion of the system.

Constraints for the common-mode feedback. The CMFB is an additional loop included in the fully-differential OTA to overcome the shortcomings discussed before. The design problem is not easy because the system should not affect the differential behavior of the OTA. Also, it should reduce drastically the common-mode signals for a very wide frequency range. Finally it should not produce additional distortion on the differential signal. A block diagram of the ideal OTA with finite common-mode gain and ideal common-mode feedback is shown in Fig.2.9.

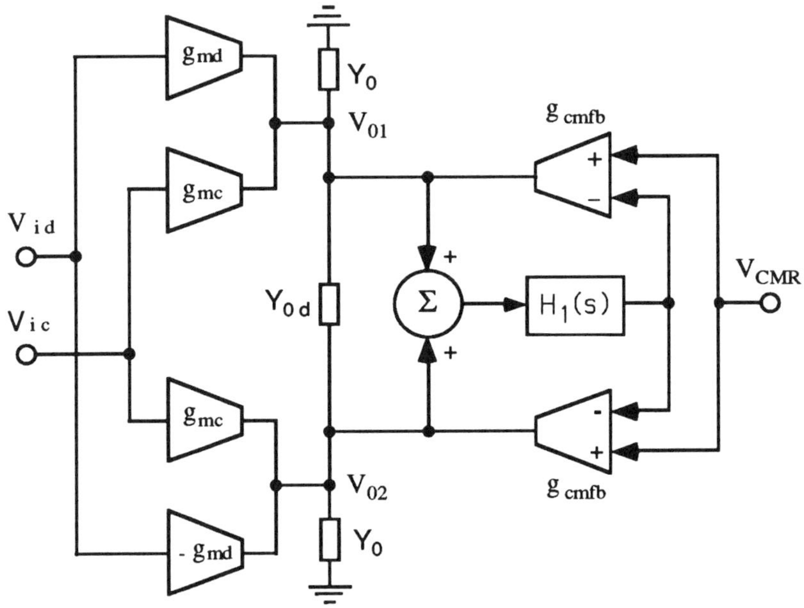

Fig.2.9. Block diagram of the OTA with an ideal common-mode feedback system.

In this diagram, The differential-mode and common-mode transconductances of the OTA are modeled by the transconductances g_m and g_{mc}, respectively. The total differential output admittance is given by $Y_{0T}=Y_{0d}+Y_0/2$. The common-mode output conductance at each output node, without the common-mode feedback, is $Y_0=g_0+sC_0$. The additional blocks are due to the common-mode feedback.

The main idea behind the CMFB is to reduce as much as possible the common-mode impedance at each output of the OTA. The common-mode detector is an ideal adder for detecting the common-mode voltage, $v_{cm}=0.5(v_{01}-v_{02})$. In Fig. 2.7, the limited frequency response of the common-mode detector is accounted in $H_1(s)$. In most of the topologies, the common-mode voltage is compared with a common-mode reference voltage, typically the arithmetic mean of the supply voltages, and the difference is fed back to the OTA output nodes. That reference voltage is represented in Fig. 2.9 by V_{CMR}. In the ideal case, the differential signal is not affected by the common-mode loop. Hence, the common-mode gain of the OTA with common-mode compensation becomes

$$A_{cc} = \frac{v_{0c}}{v_{ic}} \cong \frac{g_{mc}}{g_0 + 2\,H_1\,g_{cmfb}} \; \frac{1}{1 + s\,\dfrac{C_0}{g_0 + 2\,H_1\,g_{cmfb}}} \tag{2.41}$$

where $2H_1g_{cmfb}$ is the transconductance of the common-mode feedback loop. From this equation it is clear that increasing further the transconductance of the loop the dc common-mode gain of the OTA is further reduced. In this way a much more stable quiescent point is achieved. The other very important result of this loop is that the OTA common-mode output voltage has been strongly defined because the comparison with the common-mode reference. So, both goals are met: the definition and the stabilization of the OTA common-mode voltage. It should be noted that the common-mode impedance at the output of the OTA becomes $g_0+2H_1g_{cmfb}$ instead of g_0 and the differential output impedance is still given by $Y_{0d} + Y_0/2$.

In mixed-mode circuits the supply noise is also present at higher frequencies. So, the low-frequency behavior of the common-mode loop is quite important but also the high-frequency response of the loop. In practice, the common-mode loop introduces, at least, one additional pole. If this pole is taken into account, $H_1(s)$ can be approximated by a first-order lowpass filter, e.g. $H_1(s) \cong H_1/(1+s/\omega_c)$. In this case, for the people who are familiar with the design of filters it is clear that the common-mode impedance of the loop is a series connection of a resistor and an inductor. The value of the resistor is $1/H_1g_{cmfb}$ and the equivalent inductance is given by $1/(\omega_c H_1 g_{cmfb})$. For high frequencies, the impedance of the inductor increases resulting in higher common-mode impedance. This effect makes the CMFB less efficient at higher frequencies. Also, this equivalent inductor forms a resonant circuit with the output capacitors. In order to avoid stability problems in the loop, some design considerations must be taken into account. The common-mode loop can also be visualized as a two pole OPAMP in unity feedback configuration. In this case, the dominant pole is at the OTA outputs and the second pole is given by ω_c. If the pole of the common-mode detector is taken into account, the small-signal transfer function yields

$$A_{cc} \cong \frac{g_{cm}}{2\,H_1\,g_{cmfb}}\;\frac{1+s\dfrac{1}{\omega_c}}{1+s\dfrac{C_0}{2\,H_1\,g_{cmfb}}+s^2\dfrac{C_0}{2\,H_1\,g_{cmfb}\,\omega_c}} \qquad (2.42)$$

with $2H_1g_{cmfb} >> g_0$. This transfer function has the form of a typical biquadratic filter [1]. For dc and low frequencies, the common-mode transfer function is the same as in the ideal case, eqn. 2.41, but for high-frequencies the situation is different. For the case that $8H_1g_{cmfb} / C_0 < \omega_c$ the poles are single and positioned between $-\omega_c$ and $-2H_1g_{cmfb}/C_0$. In this case the transient response behavior is dominated by the pole with lower frequency. For the other case, $8H_1g_{cmfb}/C_0 \geq \omega_c$, the poles are complex conjugate. The frequency of the poles is given by $(2H_1g_{cmfb}\,\omega_c/C_0)^{1/2}$ and its bandwidth is given by ω_c. The transient response depends on the relationship between the real part and the imaginary part of the poles. It is very well known that the transient response of a system with complex conjugate poles placed at the left-hand side of the s plane leads to an under-damped response. The exponential decaying function of the transient response is determined by the bandwidth of the filter, in this case ω_c. The oscillating frequency depends of the frequency of the poles. This results in a very long common-mode settling time and in the generation of unwanted signals. In the general case, these signals interfere with the differential signal. Thus, even if the gain of the common-mode loop is increased further, the common-mode settling time is not decreased. The way in which the settling time of the common-mode loop can be made faster is by increasing the frequency of the pole of the common-mode loop, ω_c. Hence, in order to avoid both larger settling times and common-mode oscillations, the transconductance of the common-mode loop should be such that

$$\omega_c \geq \frac{2\,H_1\,g_{cmfb}}{C_0} \qquad (2.43)$$

Certainly, $2H_1g_{cmfb} / C_0$ plays the role of the gain-bandwidth product of the "common-mode OPAMP". This very simple equation is a quite important constraint for the design of the common-mode loop. This equation imposes a limitation to the amount of usefully common-mode feedback, due to the finite bandwidth of the common-mode loop.

The OTA with CMFB. A general and conceptual diagram of the fully-differential OTA is shown in Fig. 2.10. In this diagram, the transconductances g_{md} and g_{md}' are the differential transconductances of the input stage, which are slightly different because of the transistor mismatches. g_{mc} and g_{mc}' are defined in the same form but for the common-mode transconductances. The mismatches in the common-mode loop transconductances are included in g_{cmfb} and g_{cmfb}'. In real designs, the adder is not perfect and due to its asymmetries a small differential signal is also present at its output. This imperfection is modeled by the differential gain H_2. The load admittances are represented by Y_0, Y_0' and Y_{0d}.

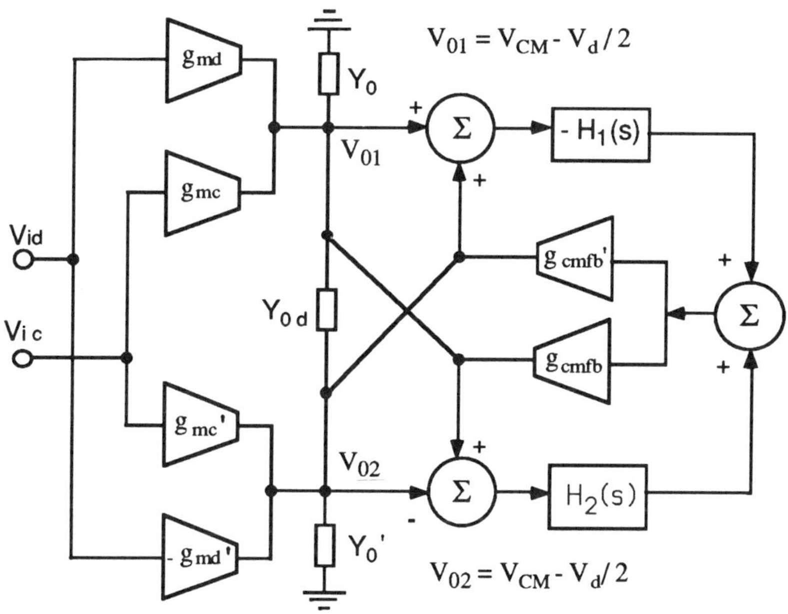

Fig. 2.10. Block diagram of the OTA including the CMFB and mismatches.

Because of the two inputs and two outputs of the OTA, four different types of transfer functions can be defined. The typical differential-mode transfer function defined as the ratio of the differential output to differential input can be deduced from Fig. 2.10 as

$$A_{dd} = \frac{v_{0d}}{v_{id}} \cong \frac{g_{mdeff}}{Y_{0d} + \frac{Y_{0eff}}{2}} \frac{1}{1 + H_2 \frac{\Delta g_{cmfb} Y_{0eff} + \Delta Y_0 g_{cmfb}}{(2 Y_{0d} + Y_{0eff})(2 H_1 g_{cmfb} + Y_{0eff})}} \tag{2.44}$$

where the subscript eff means effective value, defined as $Y_{eff} = 0.5(Y+Y')$. The factor Δ means variations, e.g. $\Delta g = g-g'$. The mismatches in the differential transconductance, g_{md} and g_{md}', and these of the grounded load admittances, Y_0 and Y_0', produce tolerances in the transfer function. The mismatches in the common-mode detector, H_2, and these of the common-mode loop transconductances, Δg_{cmfb}, and OTA output admittances, Y_0 and Y_0', introduce errors in A_{dd}, as it is evident in expression 2.44.

The other very important transfer function is the differential output to common-mode input transfer function. This transfer function can be calculated to be of the form

$$A_{dc} = \frac{v_{0d}}{v_{ic}} \cong \frac{\Delta g_{mc} + \frac{g_{mceff}(\Delta Y_0 + 2 H_1 \Delta g_{cmfb})}{Y_{0eff} + 2 H_1 g_{cmfb}}}{Y_{0d} + \frac{Y_{0eff}}{2}} \tag{2.45}$$

It can be noted that in absence of common-mode loop, this equation is similar to the common-mode gain, taken differentially at the output, already deduced in section 2.3.3. This expression can also be used for all the common-mode signals, like power-supply gains, if a properly transconductance g_{mc} is used. It has to be recognized that common-mode signals means: supply signals, even-order distortions and common-mode noise. The gain A_{dc} represents the effects of the common-mode signals on the OTA differential output signals. All those signals have to be suppressed as much as possible. The mismatches in the common-mode transconductance generate a differential current given by $\Delta g_{mc} V_{ic}$. This differential current flows throughout the differential output impedance leading to the first differential voltage component. The second voltage component is due to the mismatches in the common-mode output impedance. The common-mode input voltage is converted into a current by the effective common-mode transconductance g_{mceff}. This current is injected to the OTA outputs as a pure common-mode current. Due to mismatches in the common-mode output impedance, more current flows to ground through one side of the output than for the other side. The difference between these currents leads to a differential current flowing through the total differential output impedance. This effect is expressed in the second term of expression 2.45.

The effects of the mismatches in both common-mode transconductances and common-mode loop can be better appreciated if the CMRR is considered

$$CMRR = \frac{A_{dd}}{A_{cd}} \cong \frac{1}{\dfrac{\Delta g_{mc}}{g_{mceff}} + \dfrac{\Delta Y_0 + 2 H_1 \Delta g_{cmfb}}{Y_{0eff} + 2 H_1 g_{cmfb}}} \left(\frac{g_{mdeff}}{g_{mceff}}\right) \tag{2.46}$$

As it is evident in equation 2.46, the improvement of the CMRR is a matter of reduction of mismatches. It is important to note that the CMRR is improved by the CMFB if and only if the mismatches in the common-mode loop transconductances, Δg_{cmfb}, are small. The mismatches of the common-mode feedback loop are not due to the common-mode detector but due to mismatches in the bottom transistors M_6 in Figs. 2.4 and 2.8. These mismatches can be further reduced by placing the M_6 transistors as close as possible and using a fully-symmetrical layout. However, the CMRR is also limited by the mismatches in the grounded output admittances, ΔY_0, and Δg_{mc}.

The common-mode gain of the OTA is defined as the ratio of the common-mode output signal to common-mode input signal. From Fig. 2.10, this transfer function can be written as

$$A_{cc} = \frac{v_{0c}}{v_{ic}} \cong \frac{g_{mceff} + \dfrac{H_2 \left(\Delta g_{cmfb}\, g_{mceff} + \Delta g_{mc}\, g_{cmfb}\right)}{2 Y_{0d} + Y_{0eff}}}{Y_{0eff} + 2 H_1 g_{cmfb}} \tag{2.47}$$

and finally, the differential input signal produces also a common-mode signal given by

$$A_{cd} = \frac{v_{0c}}{v_{id}} \cong \frac{\dfrac{\Delta g_{md}}{2} + \dfrac{\left(\dfrac{\Delta Y_0}{2} + 2 H_2 g_{cmfb}\right) g_{mdeff}}{2 Y_{0d} + Y_{0eff}}}{Y_{0eff} + 2 H_1 g_{cmfb}} \tag{2.48}$$

According to relationships 2.47 and 2.48, it can be seen that both interactions are further reduced due to the CMFB. Expressions 2.44 - 2.48 will be used as guide for proper design. Experimental results of chapters 3 and 5 are in good agreement with these expressions.

2.5.3. Efficient and low-distortion CMFB. During the last few years, several implementations for the CMFB have been proposed. Some of them use a simple resistor-capacitor network for the detection of the common-mode voltage [7]. The technique is quite efficient because its power consumption is zero and its linearity is quite good. Unfortunately, the resistors reduce drastically the dc gain of the OTA and as a result the quality factor of the integrators is further reduced. Other techniques use triode biased transistors for the common-mode detection [28]. The common-mode loop transconductance is limited by these transistors, limiting further both transconductance and the bandwidth of the loop. The most common technique uses a differential pair as common-mode detector. If transistor mismatches are neglected and the ideal biquadratic law for the transistors is used, it can be shown that the voltage at the common sources of the differential pair shown in Fig. 2.1a is given by

$$v_s = v_{cm} - V_{GS} + \left(\frac{\Delta g_m}{g_m}\right)_{cmd} v_{Od} + \frac{v_{Od}^2}{8\left(V_{GS} - V_T\right)_{cmd}} \qquad \text{if } v_{Od} < 2\left(V_{GS} - V_T\right)_{cmd} \qquad (2.49)$$

where the subscript cmd means common-mode detector transistors. This equation is valid if both transistors are saturated. For low differential voltages, the ac-signal at the sources is equal to the common-mode input voltage, but for larger signals the ac-signal at the sources is affected by the differential signal. The situation is worse if the parasitic capacitors are considered. Typically, if both transistors are saturated the component proportional to v_{Od} is smaller than the second-order non-linearity. The term $(\Delta g_m/g_m)v_{Od}$ is represented in the macro-model of Fig. 2.10 by H2. In the following analysis this term is neglected.

Effects of the CMFB non-linearities. The effects of the common-mode detector non-linearities on the differential signal can be obtained using the equivalent circuit of Fig.2.10. In order to reduce the complexity of the equations, it is considered that the pole of the common-mode signal detector is at very high frequency. It is also assumed that the main distortion of the OTA is due to the CMFB and it is modeled by equation 2.49. The structure can be solved for the output voltage by typical circuit analysis, yielding

$$v_{od} = \frac{4\left(V_{GS} - V_T\right)_{cmd} y_{OdT}}{\Delta g_{cmfb}} \left[-1 + \sqrt{1 - \left(\frac{g_{md}\,\Delta g_{cmfb}}{2 y_{OdT}^{\;2}}\right) \frac{v_{id}}{\left(V_{GS} - V_T\right)_{cmd}}} \right] \qquad (2.50)$$

where the subscript 0dT means total differential output admittance. This expression can be easily expanded in a Taylor series as a function of v_{id}. From that expansion, the second and third-harmonic distortions can be computed as

$$HD2 = \frac{1}{16}\left(\frac{g_{md}}{y_{OdT}}\right)^2 \left(\frac{\Delta g_{cmfb}}{g_{md}}\right) \frac{V_{id}}{\left(V_{GS} - V_T\right)_{cmd}} \qquad (2.51)$$

$$HD3 = \frac{1}{128}\left(\frac{g_{md}}{y_{OdT}}\right)^4 \left(\frac{\Delta g_{cmfb}}{g_{md}}\right)^2 \left(\frac{V_{id}}{\left(V_{GS} - V_T\right)_{cmd}}\right)^2 \qquad (2.52)$$

In these equations, the term g_{md} / y_{OdT} is the differential gain of the circuit. For typical filter applications this is a frequency dependent gain with unity gain around the resonant frequency. The second factor is the contribution of the common-mode loop mismatches to the harmonic distortion. The common-mode feedback transconductance, g_{cmfb}, can be easily higher than the differential transconductance of the OTA by a factor of 10. Table 2.2 shows the distortions for this case and considering 1 % and 2 % transistor mismatches. For these data, $g_{md}/Y_{OdT} = 1$ has been considered.

Because the low-distortion OTA can accept input voltages with peak voltages even higher than $2(V_{GS}-V_T)_{cmd}$, the data of the last row are realistic. So, if large transistor mismatches are present in the common-mode loop, its harmonic distortion could be higher than the distortion due to the OTA input stage. The situation is worse if $v_{Od} > 2(V_{GS}-V_T)_{cmd}$. In this case the mathematical treatment becomes rather complex. This case is not discussed here. Nevertheless, the major effects are: a decrement of the differential impedance of the common-mode loop, ideally infinite, and a further increment of the distortion.

$\frac{v_{id}}{(V_{GS} - V_T)_{cmd}}$	Δ	HD2	HD3
1	0.01	- 44 dB	- 82 dB
2	0.01	- 38 dB	- 70 dB
2	0.02	- 32 dB	- 58 dB

Table 2.2. OTA harmonic distortion due to the common-mode loop non-linearities and mismatches (Calculated from 2.51 and 2.52)

An important remark on equations 2.51 and 2.52 is that both harmonic distortions are function of Δg_{cmfb}. For the fully differential OTA, see Fig. 2.8, these are indeed the mismatches of the bottom transistors M_6. Reducing as much as possible those mismatches, these type of distortions are further reduced. The linearity of the common-mode detector can be improved using source degenerated structures. However, this approach complicates the design and increase both silicon area and power consumption. A very efficient technique for low-distortion continuous-time filters that overcome the previously mentioned drawbacks is now discussed.

A low distortion approach for the common-mode loop. The design of the common-mode loop presents several design problems. Firstly, In order to improve the behavior of the OTA, both the transconductance and the bandwidth of the common-mode loop must be designed as large as possible. The unavoidable mismatches and the non-linearities of the common-mode detector produce harmonic distortion components proportional to the loop transconductance, see equations 2.51 and 2.52. Hence, in order to reduce this distortion it is desirable to reduce the loop transconductance and/or the non-linearities of the loop. From the design point of view it is, of course, more desirable to reduce only the non-linearities of the loop.

In the LDOTA, with $\beta_3=\beta_4$, the common-mode voltage is present at node v_{cm}, see Figs. 2.4 and 2.8. In fact, at low-frequencies the common-mode transfer function to v_{cm} is approximately unity. For v_{cm}, the dominant zero is located at very high frequencies and the frequency of the dominant pole is given by equation 2.27. Typically, the frequency of this pole is higher than the frequency of the differential pole. If the common-mode voltage is sensed at this node, the capacitor C_4 increases due to the sensed amplifier. This capacitance has almost no effect on the differential signals. In continuous-time filters, the output of the previous OTA is directly connected to the input of the next OTA [18-22,24,25,30]. Hence, the common-mode voltage of the previous OTA is automatically sensed at the node v_{cm} of the next OTA. This voltage can be used for the generation of the common-mode feedback of the previous OTA, avoiding the need of additional common-mode detection circuitry. As a result, the complexity of fully-differential structures is drastically reduced. Furthermore, this technique avoids the additional distortion introduced by the common-mode detector and save both silicon area and power consumption. The technique is illustrated in Fig.2.11.

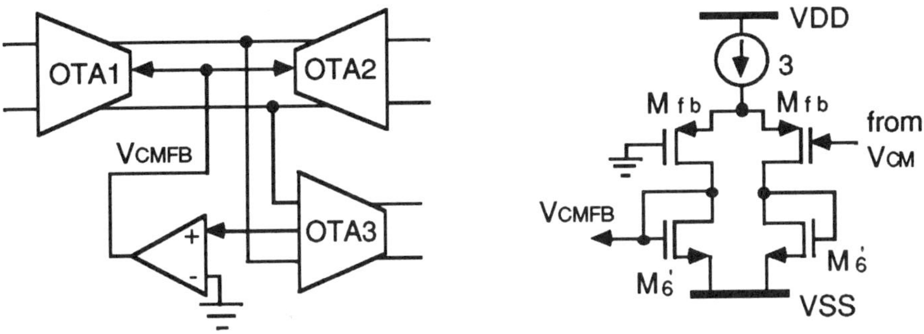

Fig. 2.11. a) Block diagram of the CMFB system. The same CMFB is used for both OTA1 and OTA2. b) Implementation of the voltage amplifier.

The sensed common-mode voltage, at the vcm node of OTA3, is compared with the common-mode reference and the error voltage is fed back to control the common-mode voltage OTA1 and OTA2. For the CMFB a single differential amplifier is only needed. Thus, both silicon area and power consumption is saved. The stability of the common-mode loop can easily be guaranteed by splitting the load capacitor in floating, if a double poly-silicon process is available, and grounded capacitors. The transconductance of the common-mode loop is determined by the transconductance of the very simple amplifier in the loop. The low distortion of the common-mode detector is automatically guaranteed because the inherently low-distortion behavior of the OTA input stage.

2.6. Conclusions.

In this chapter a systematic study of the main limitations of operational transconductance amplifiers has been carried out. The analysis includes harmonic distortion, power-supply rejection ratio, noise level and common-mode stabilization of the fully-differential structures. However, several solutions to overcome those limitations have been discussed.

The different linearization schemes are reviewed and the discussion results in a low-distortion voltage-to-current transducer. It is shown that its harmonic distortion can be reduced to the range of -60 dB for input signals higher than 2 Volts peak to peak. This topology can be used in high-frequency applications as well. The main limitations of the topology, as frequency response and rejection to supply noise, are further discussed. As a result of this, it has been shown that the design of efficient voltage-to-current transducers is feasible.

The design of fully-differential OTAs is also considered. Both the non-linearities and non-idealities of the common-mode feedback have been studied in detail. An efficient technique for the design of the CMFB that reduces the complexity of the fully-differential continuous-time filters has been proposed and discussed. These techniques are further used in the next chapters.

REFERENCES

[1] A. S. Sedra and P.O Brackett, Filter Theory and Design: Active and Passive, Matrix Publishers, Inc., Champaign Illinois, 1978.

[2] Y. Tsividis and P. Antognetti, editors, Design of MOS VLSI Circuits for Telecommunications, Prentice-Hall, Inc., Englewood Clifs, New Jersey, 1985.

[3] J. B. Hughes, "Self-Tuned RC Active Filters for VLSI," IEE Electronics letters, Vol. 22, pp. 993-994, Sept. 1986.

[4] K. S. Tan and P. R. Gray, "Fully Integrated Analog Filters Using Bipolar-JFET Technology," IEEE J. on Solid-State Circuits, Vol. SC-13, pp. 814-821, Dec. 1978.

[5] K. W. Moulding, J. R. Quartly, P. J. Rankin, R. S. Thompsom and G.A. Wilson, "Gyrator Video Filter IC with Automatic Tuning," IEEE J. on Solid-State Circuits, Vol. SC-15, pp. 963-968, Dec. 1981.

[6] D. J. Allstot and W. C. Black, Jr., "Technological Considerations for Monolithic MOS Switched-Capacitor Filtering Systems", Proc. of the IEEE, Vol. 71, No 8, pp. 967-986, Aug. 1983.

[7] M. Banu and Y. Tsividis, "Fully integrated active RC filters in MOS technology," IEEE J. Solid-State Circuits, Vol. SC-18, pp. 644-651, Dec. 1983.

[8] P. R. Gray and R. G. Meyer, " Analysis and Design of Analog Integrated Circuits," John Wiley & Sons, Singapore, 1984.

[9] Y. Tsividis, Operation and modeling of the MOS Transistor, Mc Graw-Hill, Singapore, 1988.

[10] K. Laker and W. Sansen, Design of Analog Circuits and Systems, Mc Graw-Hill, New York, to be published in 1993.

[11] D. L. Hiser and R. L. Geiger, " Impact of OTA Nonlinearities on the Performance of Continuous-Time OTA-C Bandpass Filters," IEEE Proc. ISCAS 90, New Orleans USA, pp. 1167-1170, June 1990.

[12] Y. T. Wang, F. Lu and A. A. Abidi,"A 12.5 MHz CMOS Continuous Time Bandpass Filter," IEEE Int. Solid-State Circuits Conference, New York, pp 198-199; Feb. 1989.

[13] F. Krummenacher and N. Joehl," A 4 MHz CMOS Continuous-Time Filter with On-Chip Automatic Tuning," IEEE J. of Solid-State Circuits, Vol. SC-23, pp 750-757, June 1988.

[14] R. R. Torrance, T. R. Viswanathan and J. V. Hanson," CMOS Voltage-to-current Transducers," IEEE Trans. on Circuits and Systems, Vol. CAS-32, pp 1097-1104, Nov. 1985.

[15] A. Nedungadi and T. R. Viswanathan," Design of Linear Transconductance Elements," IEEE Trans. on Circuits and Systems, Vol CAS-31, pp 891-894, October 1984.

[16] E. Seevinck and R. W. Wassenaar," A versatile CMOS Linear Transconductor/Square-Law Function Circuit," IEEE J. of Solid-State Circuits, Vol. SC-22, pp 366-377, June 1987.

[17] P. M. Van Peteghem, B. J. Haby, H. M. Fossati and G. L. Rice," A very Linear CMOS Transconductance Stage for OTA-C Filters," IEEE Custom Integrated Circuits Conference, San Diego Calif., pp 25.3.1-25.3.4, May 1989.

[18] K. Bult and H. Wallinga," A Class of Analog CMOS Circuits Based on the Square-Law Characteristic of an MOS Transistor in Saturation," IEEE J. of Solid-State Circuits, Vol. SC-22, pp 357-365, June 1987.

[19] Z. Wang and W. Guggenbuhl, " A Voltage-Controllable Linear MOS Transconductor Using Bias Offset Technique," IEEE J. of Solid-State Circuits, Vol. SC-25, pp. 315-317, Feb. 1990.

[20] Z. Czarnul, S. C. Fang and Y. Tsividis," Improving Linearity in MOS Fully-Integrated Continuous-Time Filters," IEEE Proc. ISCAS 86, Sn. Jose Calif., pp 1169-1172, June 1986.

[21] V. Gopinathan and Y. P. Tsividis," A 5V 7th-Order Elliptic Analog Filter for Digital Video Applications," IEEE Int. Solid-State Circuits Conference, San Francisco Calif., pp 208-209, Feb. 1990.

[22] H. Khorramabadi and P. R. Gray," High-frequency CMOS Continuous-Time Filters," IEEE J. of Solid-State Circuits, Vol. SC-19, pp 939-948, Dec.1984.

[23] S. L. Garverick and C. G. Sodini," Large-Signal Linearity of Scaled MOS Transistors," IEEE J. of Solid-State Circuits, Vol. SC-22, pp 282-286, April 1987.

[24] J. Silva-Martinez, M. Steyaert and W. Sansen," A high-Frequency Large-signal Very Low-Distortion Transconductor," IEEE Proc. ESSCIRC-90, Grenoble France, pp 169-172, Sept. 1990.

[25] J. Silva-Martinez, M. Steyaert and W. Sansen," A Large-Signal Very Low-Distortion Transconductor for High-Frequency Continuous-Time Filters," IEEE J. of Solid-State Circuits, Vol. SC-26, pp 946-955, July 1991.

[26] D. Senderowics, S. F. Dreyer, J. H. Huggins, C. F. Rahim and C. A. Laber," A Family of Differential NMOS Analog Circuits for a PCM Codec Filter Chip," IEEE J. of Solid-State Circuits, Vol. SC-17, pp. 1014-1023, Dec. 1981.

[27] T. C. Choi, R. T. Kaneshiro, R. W. Brodersen, P. R. Gray, W. B. Jett and M. Wilcox," High-Frequency CMOS Switched-Capacitor Filter for Communications Application," IEEE J. of Solid-State Circuits, Vol. SC-18, pp 652-663, Dec. 1983.

[28] M. Steyaert and W. Sansen, "Power-supply Rejection Ratio in Operational Transconductance Amplifiers," IEEE Trans. on Circuits and systems, Vol. CAS-37, pp. 1077-1084, Sept. 1990.

[29] Z. Y. Chang and W. M. C. Sansen, "Low-Noise Wide-Band Amplifiers in Bipolar and CMOS technologies," Kluwer academic publishers, Massachusetts: USA, 1991.

[30] M. Steyaert, J. Silva-Martinez and W. Sansen," High-Performance OTA-R-C Continuous-Time Filters," IEEE Proc ESSCIRC-91, Milano Italy, pp 5-8, Sept. 1991.

Further readings.

[31] P. R. Gray, D. A. Hodges and R. W. Brodersen, editors, Analog MOS Integrated Circuits, IEEE PRESS, New York:NY, 1980.

[32] Y. Tsividis, M. Banu and J. Khoury," Continuous-Time MOSFET-C Filters in VLSI," IEEE J. of Solid-State Circuits, Vol. SC-21, pp 15-29, February 1986.

[33] C. S. Park and R. Schawmann, " Design of a 4-MHz Analog Integrated CMOS Transconductance-C Bandpass Filter," IEEE J. of Solid-State Circuits, Vol. SC-23, pp. 987-996, Aug. 1988.

[34] P. Wu, R. Schawmann and S. Szczepanski, "A CMOS OTA with Improved Linearity Based on Current Addition," IEEE Proc. ISCAS-90, pp. 2296-2299, New Orleans, June 1990.

[35] A. Kaiser, "A Micropower CMOS Continuous-Time Low-Pass Filter," IEEE J. of Solid-State Circuits, Vol. 24, pp. 736-743, June 1989.

[36] B. Nauta, " CMOS VHF Transconductance-C Lowpass Filter," IEE Electronics letters, Vol. 26, pp. 421-422, March 1990.

[37] V. Gopinathan, Y. Tsividis, K. S. Tan and R. K. Hester, "Design Considerations for High-Frequency Continuous-Time Filters and Implementation of an Antialiasing Filter for Digital Video," IEEE J. of Solid-State Circuits, Vol. 25, pp. 1368-1378, Dec. 1990.

[38] D. Ribner and M. Copeland, "Design Techniques for Cascoded CMOS Op Amps with Improved PSRR and Common-Mode Input Range," IEEE J. of Solid-State Circuits, Vol. SC-19, pp. 919-925, Dec. 1984.

[39] A. de la Plaza and P. Morlon, "Power-Supply Rejection Ratio in Differential Switched-Capacitor Filters," IEEE J. of Solid-State Circuits, Vol. SC-19, pp. 912-918, Dec. 1984.

[40] M. Banu, J. M. Khoury and Y. Tsividis, "Fully-Differential Operational Amplifiers with Accurate Output Balancing," IEEE J. of Solid-State Circuits, Vol. 23, pp. 1410-1414, Dec. 1988.

[41] P. Van Peteghem and F. Duque-Carrillo, "A General Description of Common-Mode Feedback in Fully-Differential Amplifiers," IEEE Proc. ISCAS-90, New Orleans, pp. 3209-3212, May. 1990.

Appendix A2.1.

Harmonic distortion for the low-distortion transconductor

In this appendix the coefficients of expression 2.14 are given. The effects of the common-mode signals present in the structure are included in these coefficients. The output current of the OTA can be obtained solving a fundamental fifth order equation. In order to obtain insight in the topology, after several approximations, this equation has been reduced to a set of two second-order equations. From the solution of these equations the coefficients become

$$A_1 \cong \frac{0.75\sqrt{2\beta_1 I_B}}{1+\dfrac{\beta_1}{4\beta_2}}\left(1-\Delta_{T1}\right) \tag{A2.1}$$

$$A_2 \cong \frac{0.75\,\Delta_2}{4}\left[\frac{\beta_1\left(\eta-\Delta_{T1}\right)}{4\left(1+\dfrac{\beta_1}{4\beta_2}\right)^2}\left(1-2\,\Delta_{T1}\right)+\left(A_1\,\Delta_{T2}\right)\right] \tag{A2.2}$$

The factor Δ_2 is a result of mismatches in the structure. For $V_d = V_{GS1} - V_{T1}$ and with 2% transistor mismatches, a second-order distortion in the order of -65 dB is predicted. This value is in good agreement with the experimental results. The third coefficient is given by

$$A_3 \cong -0.75\left[\frac{\left(1+12\left(1+\dfrac{\beta_1}{4\beta_2}\right)\right)\left(\eta-\Delta_{T1}\right)\beta_1}{32\left(1+\dfrac{\beta_1}{4\beta_2}\right)^3\left(V_{GS1}-V_{T1}\right)}+\frac{A_2\,\Delta_{T2}-A_1\,\Delta_{T3}}{4}\right] \tag{A2.3}$$

The factors Δ_{Ti} are determined by the difference of the threshold voltage of transistors M_1 and M_2. These factors are computed as

$$\Delta_{T1} \cong \frac{\gamma \beta_1}{\left(\beta_1 + \beta_2\right) \sqrt{\phi_B + V_{SBO}}} \tag{A2.4}$$

$$\Delta_{T2} \cong \left(\frac{\beta_1}{\beta_1 + \beta_2}\right)^2 \frac{5\gamma}{16\left(\phi_B + V_{SBO}\right)^{1.5}} \tag{A2.5}$$

$$\Delta_{T3} \cong \left(\frac{\beta_1}{\beta_1 + \beta_2}\right)^3 \frac{13\gamma}{64\left(\phi_B + V_{SBO}\right)^{2.5}} \tag{A2.6}$$

The definition of factor γ can be found elsewhere [2, 8-10].

CHAPTER

3

DESIGN CONSIDERATIONS FOR OTA-C INTEGRATORS AND FULL CMOS RESISTORS

3.1. Introduction.

In the earlier generation of continuous-time filters the basic network elements were the inductor, the capacitor and the resistor. With the evolution of the Integrated Circuits (IC), however, the inductors have been replaced by equivalent networks composed of active devices and capacitors. The resulting topologies are called Active RC filters. The active devices of the active RC filters are either OPAMPs or OTAs, see Fig. 1.1 (chapter 1). For high-frequency applications the OTA based circuits have been preferred.

For IC filters, the integrator is the basic block. The lossless integrator consists of a voltage-to-current converter and of the integrating capacitor. The voltage-to-current converter can be implemented using a resistor and the virtual ground of an OPAMP with local feedback. The other possible implementation is using linear voltage-to-current converters, OTAs. In this text we are only interested in the second technique. In particular, in this chapter the lossless integrator and the lossy integrator and the full CMOS resistors are further discussed.

The OTA based continuous-time lossless integrator is composed by an OTA and a capacitor. In the previous chapter, the voltage-to-current converter has been discussed in detail. In this chapter, in section 3.2 the noise level of the full OTA is explored. In CMOS technologies, the capacitors are implemented as poly-silicon capacitors [1-3]. However, due to the lack of virtual grounds of the OTA based continuous-time filters, the behavior of the signal processor is affected by the parasitic capacitors. The effects of those parasitic capacitors are further discussed. In that section, it is demonstrated that the non-linearities of the bulk capacitors can produce high harmonic distortions. If the junction capacitors are comparable to the poly-silicon capacitors then these harmonic distortions are even higher than the harmonic distortions due to the OTA non-linearities. Based on the analysis of both load capacitors and OTA, the dynamic range of the lossless integrator is obtained. The linear range and noise level of the integrator are combined in the expression for the dynamic range of the integrator. From this equation, the requirements for the OTA to satisfy the filter design constraints can be found.

A couple of design examples are presented in section 3.3 [11-13]. In the first example, the design of a single-ended 150 μA/V transconductor is further discussed. Experimental results for the OTA have shown total harmonic distortions lower than -60 dB for input signals as high as 2.4 V_{pp}. These results demonstrate experimentally the low-distortion behavior predicted for the voltage-to-current transducer discussed in chapter 2. The second design is a fully-differential one. The design considerations for the common-mode loop are further discussed. Experimental results for a 6 μA/V fully-differential LDOTA show that the proposed techniques are quite efficient. For this design, the third-order distortion is lower than -65 dB for input signal up to 1. 8 V_{pp}. These results are further discussed as well.

For the design of lossy integrators, some resistors have to be used. However, for IC implementations, full CMOS resistors are required. In section 3.4, several CMOS resistors are discussed [11-13,17-30,32]. The major design considerations, the advantages and the drawbacks of the used techniques are pointed out. Some of these topologies have been used in the design of several high-performance filter realizations whose experimental results will be discussed in chapter 5. For these resistors, however, the total harmonic distortion is in the range of -60 dB for input signals up to 2 V_{pp}. For these topologies, the design considerations and the limitations as well as their advantages are further discussed in the same section. At the end of the chapter some conclusions are given.

3.2. Practical limitations of OTA based integrators.

The basic OTA-C continuous-time integrator is composed by the OTA and its associated capacitor, as shown in Fig. 3.1. The capacitors C_L and C_P correspond to the designed silicon oxide capacitor and the parasitic capacitor, respectively.

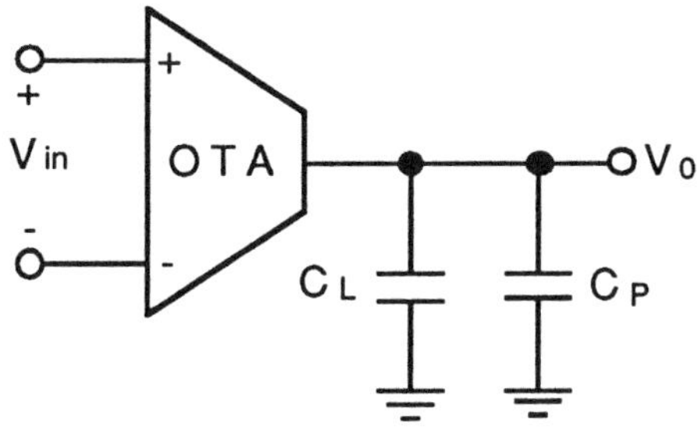

Fig. 3.1. Single-ended OTA-C integrator including parasitic capacitors.

The behavior of the integrator is determined by the characteristics of both OTA and the total load capacitor, C_L+C_P. In this section, the dynamic range of the lossless integrator is discussed. It is demonstrated that the dynamic range is mainly limited by the noise of the OTA output-stage transistors instead that for the OTA input-stage. The noise of the OTA input-stage is about 10 - 20 % of the total OTA noise. The effects on the THD of the nonlinear junction capacitors are considered as well. Also, the dynamic range of the lossless integrator using a simple differential pair input-stage and the one using the low-distortion input-stage discussed in the previous chapter are compared.

3.2.1. Noise level of the OTA. The main reason of the linearization techniques is to reduce the harmonic distortion of the OTA. So, the differential-mode range of the OTA is increased. For the linearization of the voltage-to-current transducers several additional transistors are used. These transistors increase both noise level of the OTA and power consumption. Even more, these additional transistors require some sizable silicon area. In this section we will concentrate on the study of the dynamic range of the continuous-time integrator. It is assumed that the single-ended LDOTA and the fully-differential LDOTA are the topologies already discussed in the previous chapter. For the discussion of the OTA dynamic range, the most relevant material of chapter 2 is used. The result of the section is a set of very simple equations for the description of the dynamic range of the continuous-time integrator. This set of equations allows the designer to have a very clear vision of the major limitations of the integrator dynamic range. Based on these equations, it is relatively easy to grow up to the system level. At the end of the section, the dynamic range of both simple OTA and LDOTA are compared. It is demonstrated that for the same V_{GS}-V_T the dynamic range of the LDOTA outperforms the dynamic range of the OTA by more than 34 dB. Although the power consumption of the LDOTA is higher.

Noise of the LDOTA. For the LDOTA, it has been shown that the transistors M_1-M_4 performs the voltage-to-current conversion with a transconductance given by equation 2.16. Also, in section 2.4.3, expressions for the computation of the transistor's noise factor have been carried out. It is shown that most of the noise is generated by transistors M_2, M_3 and M_7. Typically, the input referred noise of the low-distortion transducer is higher than the input referred noise of a simple differential pair by a factor 2 to 5. Nevertheless, in continuous-time filters most of the noise is generated by the transistors associated with the OTA output-stage, e.g. M_6 (the bottom transistors) and M_8 in Figs. 2.4 and 2.8. The whole noise drain current of those transistors is injected into a single output of the OTA, hence most of this current is differential. If only thermal noise is accounted, the total input referred noise of the folded cascode OTA using the low-distortion transducer is given by

$$v_{eq} = NF\, v_{eq0} = NF\sqrt{\frac{8}{3}\frac{kT}{g_m}} \tag{3.1}$$

with

$$NF = \sqrt{1 + \left(\delta + \frac{2\, g_{m7}}{g_m}\right) + \frac{2\, g_{m6}}{g_m} + \frac{2\, g_{m8}}{g_m}} \tag{3.2}$$

where NF is the noise factor of the LDOTA. The voltage v_{eqo} is the equivalent noise due to a single differential pair with transconductance equal to g_m [7-10]. In theory, v_{eqo} is the minimum achievable noise level, NF=1. It must be remembered that in the fully-differential OTA the transconductance has been defined as

$$g_m = \left[\frac{\partial\left(i_{01} - i_{02}\right)}{\partial\left(v_{i1} - v_{i2}\right)}\right]_Q \tag{3.3}$$

In the noise factor NF, relationship 3.2, the excess noise due to the transistors associated with the low-distortion transducer is given by the term between brackets. The factor δ is in the order of 2 to 4 and represents the extra noise due to the "source degeneration" technique. For an OTA using a simple differential pair as an input-stage this term vanishes. However, the noise drain current of M_6 and M_8 are present in the OTA whether the low-distortion input-stage is used or not.

In OTA based circuits, the value of the OTA transconductance g_m depends of the design constraints. Certainly, in continuous-time filters the transconductance of the OTA controls the frequency of the poles. Therefore, the OTA transconductance has a limited value. As a result, the noise is commonly dominated by the OTA output-stage transistors, M_6 and M_8 in Figs. 2.4 and 2.8. This is because the transconductance of these transistors is much higher than g_m. In continuous-time filters, however, it is common to design the OTA with $g_{m6}/g_m \approx g_{m8}/g_m \approx 10$. As a result of this the typical value for the OTA noise factor NF is around 5 to 8. It can be noted in expression 3.2 that the increment of the noise level due to the transistors used for the linearization is small, except for the noise contribution of transistors M_7.

For the whole lossless integrator the noise is the same as the OTA noise because the real poly-silicon capacitors are almost noiseless components. There are additional noise sources, due to the drain-bulk and source-bulk junctions, but these components are very small compared with those discussed before. For the case of lossy integrators, the noise contribution of the resistors, in most of the cases simulated resistors, has to be considered. The noise level of the resistors depends of its implementation. Several types of CMOS resistors will be considered in the next sections.

3.2.2. The load capacitors. The OTA based continuous-time filters use capacitors as another key element. In Fig. 3.1 it can be noted that the OTA-C integrator is implemented in open loop. Due to the lack of local feedback there is not virtual grounds in the integrator. Therefore, the output of the integrator is a high impedance node. This effect makes to the integrator sensitive to the parasitic capacitors. Those parasitic capacitors have several effects on the behavior of the integrator. This section deals with these effects, always present in the continuous-time integrator.

In the continuous-time integrator, two kind of load capacitors can be distinguish: the designed capacitor, C_L, and the parasitic capacitors represented by C_P in Fig. 3.1. Typically, the load capacitor C_L is a poly-silicon capacitor. However, the parasitic capacitors are a combination of poly and junction capacitors.

The temperature and voltage coefficients (gradients) of the poly-silicon capacitors are in the order of 10-50 ppm / °C (parts per million per Celsius degree) and 20-200 ppm / V (parts per million per applied Volt), respectively [1-3]. Because in several applications the temperature range can exceed 100 °C, variations on the absolute value of the capacitance of about 0.5 % can be expected. The effects of the voltage coefficient are less important. This is because for typical applications the swing of the signal is lower than 2.5 Volts. This leads to capacitor variations of the order of 0.05%. Then, these effects can be neglected. The effects of the temperature variations can be minimized with an automatic tuning system, to be considered in the next chapter.

There are two types of parasitic capacitors, namely, gate and junction capacitors. These capacitors are due to the input- and output-stages of the OTAs connected to the output of the integrator. Also the top and bottom parasitic capacitors due to the poly-silicon capacitor have to be considered. The gate-source, gate -drain and top capacitors are poly-silicon capacitors and so they are quite linear. Strictly speaking, due to the modulation of the channel, the gate-source and gate-drain capacitors are voltage dependent capacitors. However, if the transistor is biased in deep saturation, the channel modulation is small and so they can be approximated as linear capacitors. Under these conditions, the effect of these poly-silicon capacitors is an increment in the load capacitance. These capacitors can be considered as merged with the load capacitor and leading to larger load capacitor tolerances. Their value can be easily computed and they have to be incalculated in the load capacitor. Unfortunately, the junction capacitors are non-linear voltage dependent capacitors. The effects of these capacitors are twofold. Firstly, there is an operation point dependent component which increases the load capacitance just like the effect of the gate capacitors. The value of this capacitor is not very well controlled. The second effect of these capacitors is an increment of the harmonic distortion of the integrator. A detailed analysis of the effects of these non-linearities is carried out in this section. Also, a relationship between the poly-silicon capacitor and the non-linear capacitors as function of the THD is derived. From this relation, a rule of thumb for the computation of the minimum poly capacitor needed for the design of low-distortion integrators is obtained.

Effects of the nonlinear junction capacitors. In CMOS technologies, the typical voltage dependence of a junction capacitor has the form [4, 6, 7]

$$C_j = \frac{A_J\, C_{j0}}{\sqrt{1 + \dfrac{V_{DB} + v_0}{\phi_j}}} = \frac{A_J\, C_{j0}}{\sqrt{1 + \dfrac{V_{DB}}{\phi_j}}\,\sqrt{1 + \dfrac{v_0}{V_{DB} + \phi_j}}} \tag{3.4}$$

with A_j, C_{j0}, V_{DB} and ϕ_j being the junction area, the junction capacitance per unit of junction area with zero biasing, the junction bias voltage and the surface potential, respectively. v_0 is the ac capacitor voltage, see Fig. 3.1. For typical CMOS processes C_{j0} and ϕ_j are in the order of 0.3 fF/μm^2 (10^{-15} Farads per junction area in squared microns) and 0.6 Volts, respectively. A Taylor series expansion of equation 3.4 allows to compute the harmonic distortion introduced in the integrator by the junction capacitor. If the Taylor series expansion of equation 3.4 is used for the characterization of the junction capacitors, the output voltage of the continuous-time integrator shown in Fig. 3.1 can be approximated by

$$v_0 \cong \frac{Q_{in}}{C_L'}\left(1 + \left(\frac{C_j'}{2\left(C_L + C_j'\right)}\right)\frac{Q_{in}}{C_L' V_B'} - \left(\frac{C_j'\left(3\,C_L + C_j'\right)}{8\left(C_L + C_j'\right)^2}\right)\left(\frac{Q_{in}}{C_L' V_B'}\right)^2 + ..\right) \qquad (3.5)$$

where Q_{in} is the OTA injected charge. This charge is equal to the time integral of the OTA output current. For a single sinusoidal input, the peak magnitude of this charge is equal to $g_m V_{in} / \omega_0$, with V_{in} and ω_0 the amplitude and frequency of the input signal, respectively. The voltage V_B' is the "effective bias voltage" of the junction capacitor and defined as $V_{DB} + \phi_j$. For applications using supply voltages of ± 2.5 Volts, V_B' is around 3 Volts. The capacitor C_j' is defined as equal to C_j evaluated at $v_0 = 0$. In addition, $C_L' = C_L + C_j'$. Therefore, C_L' is the real load capacitor for the OTA, as it can be noted in 3.5.

From expression 3.5 and the definitions of HD2 and HD3, see section 2.2, the harmonic distortions due to the non-linear capacitors can be easily computed. If this is done, it can be shown that HD2 and HD3 are proportional to $Q_{in}/(C_L' V_B')$ and $(Q_{in}/(C_L' V_B'))^2$, respectively. It can be seen that the harmonic distortions are higher at low frequencies, $Q_{in} \approx g_m V_{in} / \omega_0$. This is because a single integrator has been considered. Hence, the low-frequency gain of the integrator is very high, in the ideal case infinite at dc. In practical applications the integrator is connected in a feedback loop configuration. The feedback reduces the low-frequency gain of the integrator and so it reduces further the distortion. Hence, it is more realistic to compute the distortion at the unity gain frequency of the integrator, $\omega_0 = g_m / C_L'$. Theoretical harmonic distortions for the case $V_0 = 0.5$ V ($v_0 = 1\ V_{pp}$) and $V_B' = 3$ V are shown in table 3.1.

$\frac{C_L}{C_j'}$	HD2 (dB)	HD3 (dB)
1	-33.6	- 61.2
2	-37.2	- 63.4
4	-41.6	- 66.9

Table 3.1. Harmonic distortions due to the non-linearities of the junction capacitors.

From this table it is clear that the most important effect of the load capacitor non-linearities is a further increment of the second-order harmonic distortion, HD2. This is a major problem in single-ended topologies because this distortion appears just at the output of the integrator. As a result, in single-ended topologies this second-order distortion can be larger than the HD2 due to the OTA non-linearities. Fortunately, this effect is not severe in fully-differential structures. In the case of fully-differential OTAs, the even-order non-linearities are further rejected due to the symmetry of the circuit and because the voltage subtraction takes place just at the output of the integrator. Furthermore, the even order distortions are common-mode signals. Then, they are reduced at the OTA output because the low common-mode impedance of the common-mode feedback.

Form table 3.1, it can be noted that for junction capacitors comparable to the poly capacitor the HD3 due to the non-linear capacitors is comparable to the HD3 due to the OTA ($\approx$ -60 dB). However, it should be recognized that the sidewall capacitors are always present. These are nonlinear capacitors as well. The sidewall capacitors have an output voltage dependency similar to the relationship 3.4 but with a cubic root function rather than a square root behavior [6,7]. Because this fact, their effects lead to smaller distortion contributions. When both bottom and sidewall capacitors are taken into account in table 3 the HD3 increases approximately by 1 or 2 dB.

According to table 3.1, for low-distortion applications, the minimum poly-silicon capacitance needed in the integrator can be calculated by the following rule of thumb: The total junction capacitor at the output of the OTA, C_j', must be lower than 0.25 C_L, C_L being the poly-silicon capacitor. For instance, in high-frequency applications the total junction capacitor could be of the order of 1 pF. Hence the poly-silicon capacitors have to be larger than 4 pF. In the case of several OTAs connected to the same node, the total junction capacitor have to be used in this rule. If floating poly-silicon capacitors are used, then the bottom capacitors must be taken into account as well.

A last but important remark on expression 3.2 and table 3.1 is: These data have been calculated assuming that the bias voltage for the junction capacitor is 2.5 V. In the OTA, this means that the bulk of all the OTA output-stage transistors have been connected to the supply rails. In the case of cascode output-stages using transistors with their bulk connected to their own source, the bias voltage of the junction is lower. As a result, larger amounts of distortion can be obtained.

3.2.3. Dynamic range of the LDOTA. In the previous chapter, the harmonic distortion for the LDOTA has been computed. For the LDOTA the distortion is mainly due to the voltage-to-current transducer. Also, in section 3.2.2., the effects of the nonlinear capacitors have been studied. In the lossless integrator both effects are present at the same time. Because the nonlinear effects of the capacitors are function of the frequency, the computation of the harmonic distortions becomes quite complex and rather involved, even for very simple circuits. In these cases the computation of the harmonic distortions requires Volterra series expansion. These techniques are not discussed here, but mathematical and circuit treatment of these problems can be found in references 4 and 5. From the designer point of view, it is interesting to learn that the capacitive non-linearities can be further reduced by properly dimensioning the capacitors, see section 3.2.2. The computation of the harmonic distortion components is further simplified if the capacitors are considered as linear elements. This criterion is used through the whole text. Hence, it is considered that the dominant non-linearities are those due to the input-stage of the LDOTA.

In section 2.3, it has been shown that the linear range of the LDOTA is, approximately, given by 1.41 $(V_{GS1}-V_T)$ $[V_{RMS}]$. If only thermal noise is considered, the noise level of the transconductor is given by relationships 3.1 and 3.2. Hence, the dynamic range (the ratio of the OTA linear range, RMS value, to the input referred integrated noise) of the lossless integrator can be computed as

$$DR \cong \frac{1.4\ (V_{GS1} - V_T)}{NF\ v_{eq0}\sqrt{\Delta f}} = \frac{1.4 \times 10^{10}\ (V_{GS1} - V_T)\sqrt{g_m}}{NF\sqrt{\Delta f}} \tag{3.6}$$

with Δf the frequency range of interest. This very simple expression together with equation 3.2 are the most powerful tools for the design of LDOTAs and lossless integrators. These relationships show clearly the major practical limitations and the major characteristics of the most important parameter of the LDOTA: the dynamic range. In the particular case of low-pass filters the frequency range Δf is almost equal to g_m/C_L. Therefore, the dynamic range is proportional to the square root of the load capacitor used. Hence, the larger the load capacitor C_L the higher the dynamic range is. However, for larger capacitors used larger transconductances have to be used as well. As a result, both power drain and silicon area increases, and in some cases the noise factor increases too.

LDOTA against OTA. The harmonic distortion and the noise level are the limits for the OTA dynamic range. Although the LDOTA reduces further the distortion, it is not sufficient to improve the linearity of the OTA if the dynamic range is not increased. If an OTA using a simple differential pair input-stage is compared with a LDOTA, for the same transconductance, the input referred noise of the LDOTA is higher than the former only by a factor of 1.1 to 1.5. The noise behavior is dominated by the output-stage of the OTA. Hence, the noise factor of the LDOTA is higher than the noise level of the OTA in the very worst case by 6 dB. For this case and for the same transconductance and for the same harmonic distortion, from equations 2.8 and 2.9, the ratio of the dynamic range for both transconductors becomes

$$\frac{DR_{LDOTA}}{DR_{OTA}} \cong 4\left(1+\frac{\beta_1}{4\,\beta_2}\right)^2 \tag{3.7}$$

For the case of $\beta_1/\beta_2 = 10$, the dynamic range of the LDOTA is 50 times, 34 dB, higher than the dynamic range of the simple OTA. However, it should be recognized that these benefits cost. The bias current of the input-stage of the LDOTA is around 10 times higher than the input-stage bias current needed for a simple OTA. This does not necessarily mean that the power consumption of the LDOTA is 10 times higher than this of the OTA power consumption but the drain current of the LDOTA is certainly higher. From equation 2.9, if the bias current of the OTA input-stage is increased by a factor 10 then its distortion decreases by 20 dB. Then the difference in dynamic range is only 14 dB. This difference is due to the cross-coupling technique used in the LDOTA (7 dB of reduction in HD3) and also due to the internal feedback of the active source degeneration (≈ 6 dB). However, increasing the OTA bias current by a factor 10 the effective gate-source voltage, $V_{GS1} - V_T$, increases by a factor of 10. For low supply voltages, this fact makes this approach unsuitable.

3.3. Design of single-ended and fully-differential integrators.

In order to illustrate the design procedure two integrators are considered in this section. Hand calculations, simulations and experimental results for both single-ended and fully-differential OTAs are compared. Firstly, the design procedure of a single-ended LDOTA is considered. Afterwards, the design of a 6 μA/V fully-differential LDOTA is discussed. The LDOTAs together with the load capacitors determine the continuous-time lossless integrators. These integrators are used in the design of several higher order filters whose experimental results are discussed in chapter 5.

3.3.1. A 150 μA/V single-ended LDOTA: A design example. For the single-ended design, a 150 μA/V LDOTA is considered. In this OTA, a CMOS 3 μm N-well process has been used. The process parameters are given in appendix A1.1. For the design of a programmable second-order filter, to be discussed in chapter 5, a LDOTA is required with a dynamic range larger or equal to 85 dB for a frequency ranging from 1.5 MHz up to 2.5 MHz [11]. The supply voltages are only ± 2.5 V.

Design procedure. Because the high-frequency range, the flicker noise can be neglected. Then, the equations 3.4, 3.5 and 3.6 can be used. Because the very high dynamic range required, the maximum bias voltage for the input-stage is used, $V_{GS1} - V_T = 0.6$ Volts. Therefore, the common-mode range of the LDOTA must be higher than 0.6 Volts. In this design, a common-mode range of 0.7 V is employed. The common-mode range can be increased by increasing the transconductance of the transistors M_7, see Fig. 2.4. The drawback is that the noise factor increases too. For this biasing, it is expected to achieve a linear range for the differential input signal of the order of 2.4 Vpp.

Since the transconductance, the passband frequency range, the bias voltage of the OTA input-stage and the dynamic range are known, the minimum OTA noise factor NF can be computed from relationship 3.6 as 8. This noise factor imposes some limitations to the output-stage transistors of the OTA. Before computing the dimensions of these transistors, the transistors of the low-distortion input-stage are calculated.

Because the very low-distortion required and in order to increase further both linearity and input range, $\beta_1/\beta_2 = 10$ is used. With this value and from equation 2.17b, HD3 should be about -64 dB for an input signal of 1.2 V_{pp} ($V_d = V_{GS1} - V_T = 0.6V$). The additional feedback of the voltage-to-current transducer increases this limit by a factor 2. The transconductance of M_1 can be obtained from the expression 2.16. If the second-order effects are neglected, g_{m1} is obtained as 700 μA/V. The bias current I_B can be obtained from the quadratic relationship between the drain current and $V_{GS1}-V_T$ as

$$I_B = \frac{\beta_1}{2}(V_{GS1} - V_T)^2 \tag{3.8}$$

When numerical values are inserted in this equation, a current of 210 μA is obtained. With this value and because the transconductance of M_1 is known, $W_1/L_1 = 23.3$ is computed. In order to reduce the effects of the transistor output conductance, a minimum gate length of 6 μm is employed. From these results, the whole dimensions of the other transistors can be calculated; $W_1 = 23.3 \times 6\ \mu m = 140\ \mu m$, $W_3 = W_4 = 0.5\ W_1 = 70\ \mu m$ and, if $L_2 = L_1$, then $W_2 = W_1 / 10 = 14\ \mu m$. Actually, due to the bulk effects, a reduction in the transconductance in the order of 10 % is expected. It is desirable to increase both bias current and transistor dimensions in the order of 10 % for the compensation of this effect.

Because the main goal of this design is to demonstrate the behavior of the voltage-to-current transducer, the high-frequency poles of the cascode transistors are placed at very high-frequency. The same is done for the parasitic poles due to the differential to single current converter. The only constraint for transistors M_6, M_7 and M_8 is to keep the noise factor lower than 8. The final dimensions and bias currents of the whole transistors are given in table 3.2.

For these dimensions a hand calculated noise factor of 8.4 is obtained. The effect of this slightly larger value of NF, calculated as < 8, is a reduction of 0.3 dB in the OTA dynamic range. On the other hand, simulated results have shown parasitic pole frequencies higher than 250 MHz.

transistor	W/L (μm/μm)	current (μA)
M1	150/6	220
M2	15/6	0
M3,M4	75/6	110
M5	500/3	330
M6	250/3	330
M7	350/6	220
M8	500/4.5	660

Table 3.2. Dimension and bias current of the transistors of the single-ended low-distortion OTA.

Simulated and experimental results. Some of the most important experimental and simulated results (Spice level 2) are given in table 3.3. For the simulated results, 1 % transistor mismatches have been included.

From this table it can be seen that the experimental transconductance is larger than the simulated one. This means that the experimental gate source voltage was lower than simulated. In fact, in simulations, the THD is lower than -60 dB for input voltages up to 2.8 Vpp, but in experimental measurements this range was only 2.4 Vpp. Also, it should be recognized that for these simulations Spice2G6 level II was used. The limitations in the transistor model of this version are very well known [4, 6, 7, 33]. Nevertheless, the experimental results fit well with the expected and simulated ones.

parameter	simulated	experimental
dc gain	50 dB	45 dB
transconductance	150 μA/V	155 μA/V
THD @ 2.4 Vpp	- 62 dB	- 60 dB
dynamic range	88 dB	89 dB
CMRR @ 1.5 MHz	52 dB	48 dB
PSRR+ @ 1.5 MHz	42 dB	40 dB
PSRR- @ 1.5 MHz	35 dB	30 dB

Table 3.3. LDOTA simulated and experimental results.

The measured harmonic distortion components are shown in Fig. 3.2. The second harmonic distortion HD2 dominates at low voltages. The reasons are twofold. Firstly, the mismatches in the transconductor and the junction capacitors produce a small second-order distortion. The mismatches can not be completely eliminated but they can be reduced by layout techniques. The effect of the junction capacitors can be reduced by using smaller dimensions for the output-stage transistors and optimizing the layout for minimal drain area. The second reason of the large HD2 is due to the differential to single-ended conversion, transistors M_6 in Fig. 2.4, as it has been discussed in chapter 2. The additional poles and the mismatches present in one path of the current avoids the exact cancellation of the even order harmonic distortion components.

The measured third-harmonic distortion fits very well with the theoretical and simulated results. These results show that the technique allows to design very low distortion transconductors, HD3 < -65 dB, for differential input voltages up to 2.4 Vpp. This range represents almost 50 % of the total supply voltages and around 41 % of the total transducer bias current. The spectrum of the OTA output for a 1.2 V_{pp} single input signal is shown in Fig. 3.3. For single input signals the low distortion input range is reduced almost by a factor of 2. The main reason is the limited common-mode range of the structure. It has to be remembered that the common-mode range of the LDOTA is only 0.7 Volts.

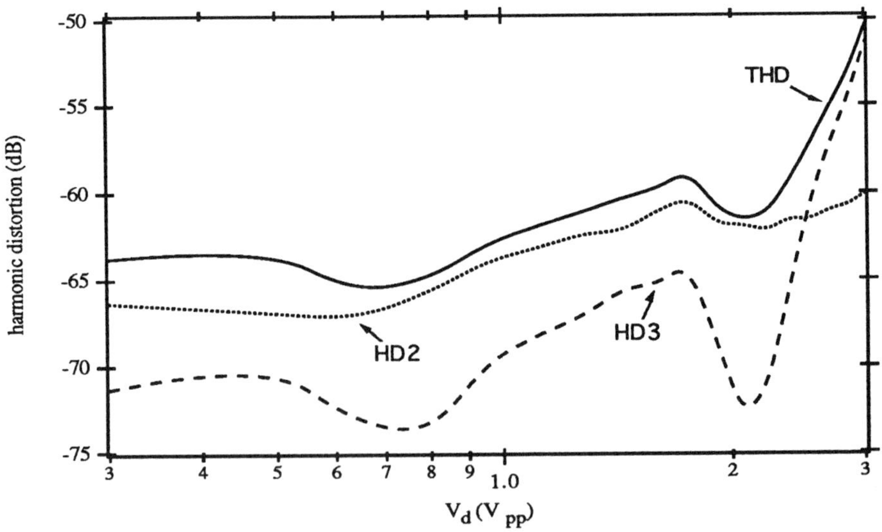

Fig. 3.2. Measured OTA Harmonic Distortion with a fully-differential input signal. The THD is lower than - 58 dB for $V_d < 2.4$ V_{pp}.

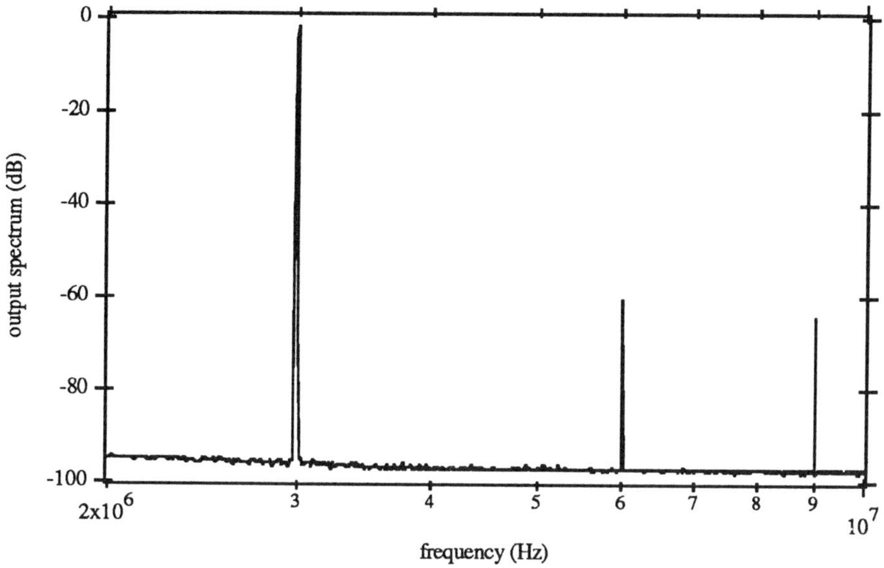

Fig. 3.3. Spectrum of the OTA output voltage for a single input signal. The input voltage is 1.2 V_{pp}.

3.3.2. A Fully-differential LDOTA: design example. The design procedure for the fully-differential OTA input-stage is similar to the procedure followed for the single-ended transconductor. In this section, the design is focused on the common-mode feedback system.

Design of the LDOTA. For the design of a fourth-order 100 kHz low-pass filter, to be discussed in chapter 5, a 6 μA/V fully-differential OTA is required. Due to filter constraints, the load capacitors are in the range of 8 - 14 pF. The power consumption has to be as low as possible, and the dynamic range as high as possible. Using the LDOTA, the following design constraints are used: $\beta_1/\beta_2=10$ and $V_{GS1}-V_T = 0.5$ Volts. Following the same approach as in the previous example, both bias currents and transistor dimensions can be obtained. In this case, an other very important design constraint to take into account is the frequency of the common-mode pole. It must be remembered that the gain-bandwidth product of the common-mode feedback is limited by this pole (see equations 2.42 and 2.43 of section 2.5). More details of the CMFB will be discussed shortly. The bias current of the low-distortion transducer, I_B, has been calculated as 6 μA. For transistors M_1, M_3 and M_4, gate lengths of 6 μm are employed. This is in order to reduce the effects of the transistor output conductance's. Because for M_2, W/L =1/10, a transistor width of 3.6 μm has been used. Although the small width of the transistor, the flicker noise behavior of the transistor is not affected because the bias current of the M_2 transistors is zero. Therefore, this width value is used in order to decrease the parasitic capacitors.

The second pole of the LDOTA is determined by transistors M_5, see Fig. 2.8. Using a bias current of $3I_B$ = 18 μA for M_8, the bias current flowing through M_5 is 9 μA. For this transistor, with W/L equal to 24 μm/ 4.8 μm, the frequency of the parasitic pole is higher than 12 MHz, 120 times higher than the cutoff frequency of the filter. For transistors M_6, the major design constraints are associated with their parasitic poles and the noise considerations. There are 2 parasitic poles associated with these transistors, see Fig. 2.8. The first one is at the node named V_{CMFB} , at the gate of the bottom transistor. The other one is associated with the drain of the same bottom transistor M_6. These poles are in the common-mode feedback loop and their frequencies have to be at least 4 times higher than the gain-bandwidth product of the loop. Otherwise, these poles reduce further the phase margin of the common-mode feedback loop. For low-frequency applications it is not difficult to satisfy this constraint because the transistors M_6 are N type and their bias current is high enough. For instance, for $(W/L)_6$ = 12 μm / 6 μm and using a bias current of 9 μA, the frequency of its parasitic pole is higher than 10 MHz. For this application, this frequency is high enough because the gain-bandwidth of both differential-mode loop and common-mode loop are lower than 2 MHz. Nevertheless, this is not the only constraint associated with these transistors. For low-frequency applications the noise behavior is dominated by the flicker noise rather than by the thermal noise. For typical N-well processes, the flicker noise coefficient, K_f, of the NMOS transistors is much higher than for the PMOS transistors. For the 1.2 μm process used $K_{fN} = 24\ K_{fP}$. Therefore, for low-frequency the major noise contributions are due to the N transistors. Typically, the transistors M_1, M_3 and M_4 have lower noise contributions than transistors M_6 and M_7. The major reasons for that are due to their larger bias current and their larger transconductances. However, the noise drain current of these transistors is almost differential noise. Hence, when the noise current is referred to the input of the OTA their noise contribution is amplified by the ratio of the transistor transconductance to the OTA transconductance, see expression 3.2. Because for transistors M_6 and M_7 the noise drain current is important rather than the input referred noise, it is desirable to reduce their bias currents and to increase as much as possible the gate length of the transistors [7-10]. It is obvious that this approach reduces further the frequency of the parasitic poles associated with M_6. In the case of M_7, this approach reduces the common-mode range of the OTA. As a result, there is a trade off among these parameters. For M_6 a gate length of 4.8 μm has been used and for transistors M_7 a gate length of 6 μm has been employed. The width of M_7 has been determined from the common-mode range constraint. For $(W/L)_7 = 1.5$, a common-mode range of 0.7 Volts is obtained.

The common-mode feedback. In order to increase the gain-bandwidth product of the common-mode loop, the frequency of the dominant pole should be as high as possible, typically higher than the gain bandwidth of the differential-mode loop. For the dimensions of the transistors associated with the voltage-to-current transducer, a common-mode pole at v_{cm} of around 1.2 MHz has been obtained from simulations. This pole frequency is enough for the 100 kHz low-pass filter intended to design. The gain-bandwidth product of the common-mode loop is determined by the gain of the common-mode voltage V_{CMFB}, see Fig 2.10 of chapter 2. This gain-bandwidth product is approximately given by

$$GBW_{cm} = \left(\frac{g_{m6}}{g_{m6}'}\right)\frac{g_{cmfb}}{C_0} \tag{3.9}$$

Typically, but not necessarily, $g_{m6}=g_{m6}'$. The gain-bandwidth product of the common-mode loop is controlled by the grounded capacitor C_0 and the transconductance g_{mfb} of the P transistors M_{fb}. In this design, the bias current of the common-mode feedback block is equal to the bias current of M_8, $3I_B=18$ μm. For this design, $g_{m6}=g_{m6}'$ and $g_{cmfb}=g_{mfb}/2 = 3\ g_m = 18$ μA/V. So, under maximum load conditions, the gain-bandwidth of the common-mode loop is higher than the cutoff frequency of the filter approximately by a factor of 2. Under minimum load conditions, the gain-bandwidth product is around 450 kHz, leading to a phase margin of 65 degrees for the common-mode loop. However, in the filter more than 1 output-stage is connected to the common-mode feedback loop. In fact, in fully-differential filters, at least 2 OTAs are connected at the output, one for the direct path and other one for the feedback. An exception is the termination of some ladder filters. However, in this case a simulated resistor is present and very often this resistor is simulated with another OTA. Because there are 2 output-stages of the OTAs connected to the CMFB, the transconductance of the loop increases by a factor of 2. As a result, the gain-bandwidth product of the common-mode loop is in the range of 400 kHz and 900 kHz for maximum and minimum load capacitors, respectively. The phase margin is in the range of 48 and 65 degrees for minimum and maximum load capacitors, respectively. The dimensions and bias currents of the transistors are shown in table 3.4.

transistor	W/L (μm/μm)	I_{BIAS} (μA)
M1	6/6	6
M2	3.6/36	0
M3,M4	12/6	3
M5	24/4.8	9
M6, M6'	12/4.8	9
M7	9/6	6
M8	36/4.8	18
M9	12/6	6
Mfb	24/4.8	9

Table 3.4. Dimension and bias conditions for the transistors of the fully-differential LDOTA.

For this design, if only thermal noise is taken into account, the noise factor of the LDOTA is about 7. The LDOTA dynamic range is calculated in the order of 79 dB. If the flicker noise is included, the noise increases by 2 or 3 dB. The power consumption of the LDOTA is in the order of 210 μWatts.

Experimental Results. The fully-differential LDOTA was fabricated in a 1.2 μm N-well single poly CMOS process. The measured HD3 component and the total harmonic distortions are shown in Fig. 3.4. The third-harmonic distortion is lower than -65 dB for fully-differential input voltages up to 1.8 Vpp and lower than -60 dB for V_{in} < 2Vpp. The harmonic distortion increases rapidly for higher voltages. This is because the swing of the input signal is of the order or even higher than the common-mode voltage of the input-stage. For low input voltages, a second-order distortion slightly smaller than HD3 is present. This component increases the THD to the order of -63 dB. It is expected that improving the symmetry of the layout the even-order distortions can be reduced by 2 or 3 dB. In this plot, only the second and the third-harmonic distortions have been accounted. For input signals smaller than 2 Vpp the higher order distortions were hidden in the noise. These distortions are present for input signals in the order of 2.2 Vpp. For this voltage, the HD5 component is around -60 dB and for V_{in} = 3 Vpp is of the same order as HD3. This is mainly due to the limited common-mode range of the transconductor. Certainly, at those very large input voltages the higher-order harmonic distortions of the voltage-to-current transducer are considerable.

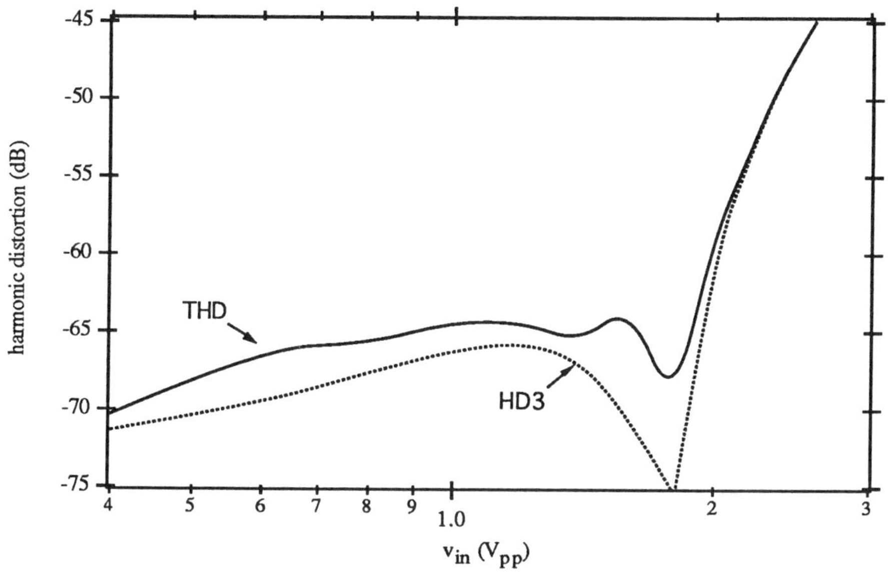

Fig. 3.4. Measured harmonic distortions for the fully-differential OTA

In order to test the behavior of the common-mode feedback loop, a load impedance consisting of a 150 kΩ floating resistor and two grounded capacitors have been used, as shown in Fig. 3.6. Hence, at low frequencies the common-mode gain of the LDOTA is given by the ratio of the common-mode transconductance of the OTA to the common-mode feedback transconductance. The dc value and the dominant zero of the OTA common-mode transconductance are determined by g_{07} and g_{07}/C_{P7}, respectively. The g_{07} and C_{P7} are the grounded conductance and the total parasitic capacitance associated with the drain of M_7 (see Fig. 2.8). For this design, the frequency of the zero is around 350 kHz. For the common-mode feedback loop, the dominant pole is given expression 3.9. The experimental results for 2 different load capacitors are shown in Fig. 3.5. For the case C_0 = 10 pF, the dominant pole of the common-mode gain is located at 280 kHz. For C_0 = 8 pF, the pole is around 350 kHz. This frequency is 3.5 times higher than the filter cutoff frequency. The phase response of the common-mode loop is good enough because the second pole frequency is around 1.2 MHz. Using the same setup and C_0 = 10 pF, the experimental results of table 3.5 have been obtained.

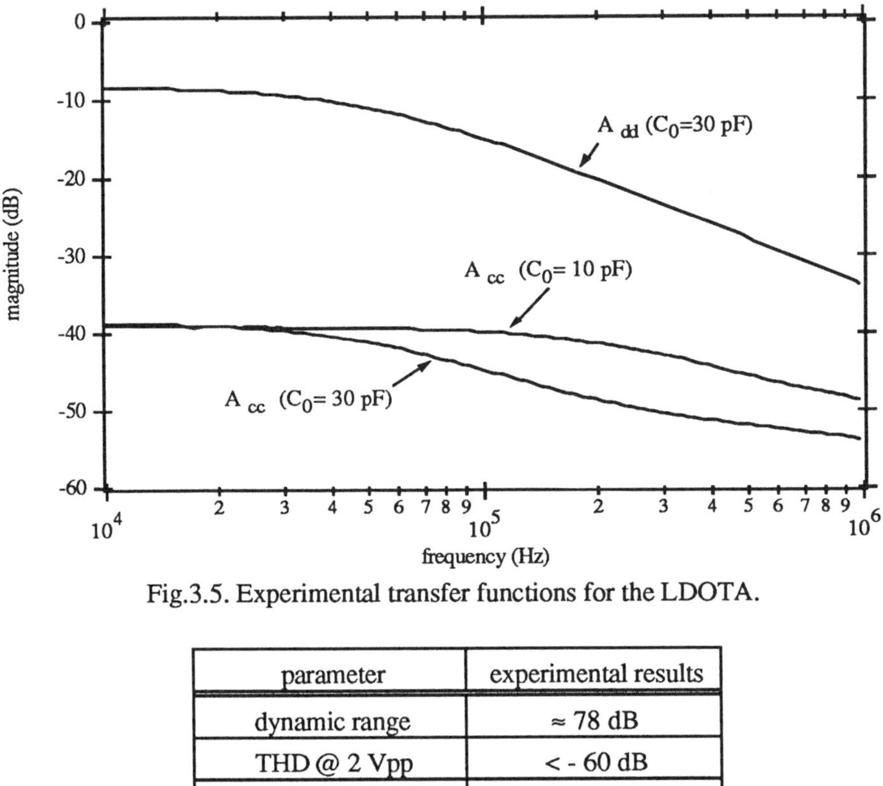

Fig.3.5. Experimental transfer functions for the LDOTA.

parameter	experimental results
dynamic range	≈ 78 dB
THD @ 2 Vpp	< - 60 dB
CMRR	@ 20 kHz 70 dB @ 100 kHz 55 dB
PSRR, VDD	@ 20 kHz 58 dB @ 100 kHz 52 dB
PSRR, VSS	@ 20 kHz 55 dB @ 100 kHz 51 dB
common-mode offset	≈ 45 mV
power consumption	210 μWatts

Table 3.5. Measured results for the low-distortion fully-differential OTA.

3.4. The lossy integrator.

In the previous section, the lossless integrator has been discussed in great detail. Ideally, the pole of a lossless integrator is on the origin of the s plane. These integrators connected in series lead to unstable systems, due to both the very large dc gain of the OTA and the lack of feedback. On the other hand, if two or more lossless integrators are connected in a loop, with negative loop gain, the poles can be located on the real axis or on the imaginary axis or/and complex conjugate poles at both sides of the s plane. For the case of 2 integrators the system becomes a resonator. A resonator is a second-order system with complex conjugate poles placed on the jω axis. The impulse response of the resonator is a sinusoidal function. In this case the circuit becomes an oscillator.

In general, the lossless integrators are not sufficient for the design of filters. The shape of the transfer function is related to the real part of the system poles. Hence, for the control of the shape of the filter, it is necessary to include certain circuit elements that introduce a controllable and negative real part in the poles of the transfer function [14-17]. The mostly used method is controlling the dc gain of some integrators of the system. The integrator with finite dc gain is called the lossy integrator [16-17]. Its dc gain can be controlled by a resistor as shown in Fig. 3.6.

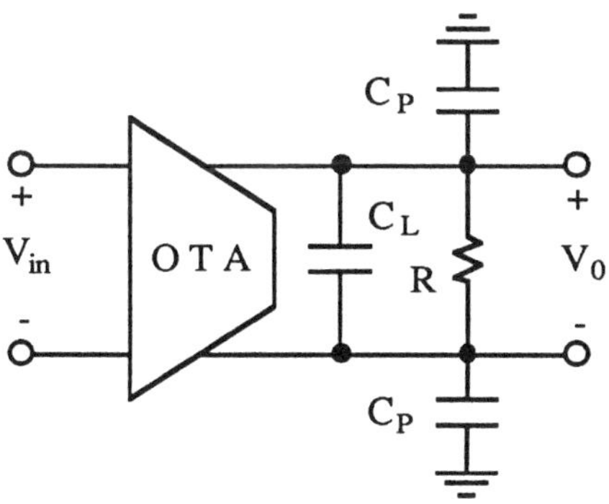

Fig. 3.6. The continuous-time lossy integrator.

The ideal transfer function of the lossy integrator shown in Fig. 3.6 is given by:

$$\frac{v_O}{v_{in}} = \frac{1}{\frac{1}{g_m R} + s\frac{C}{g_m}} = g_m R \frac{1}{1+sRC} \qquad (3.10a)$$

with $C=C_L+C_P/2$. This transfer function corresponds to the typical first-order low-pass filter with dc gain equal to g_mR and pole frequency at $\omega = 1/RC$.

The resistor R can be implemented by using a poly-silicon resistor or a diffused resistor. These passive resistors are quite linear and can be fabricated in a standard CMOS process. However, in order to tune the real part of the poles an adjustable resistor has to be used. The main disadvantage of the passive resistors is the lack of tunability.

The remaining of this chapter deals with the different implementations of full CMOS tunable resistors. For continuous-time applications, the most important parameters of the CMOS resistors are linearity and tuning range. In this section, the most relevant characteristics of several implementations are pointed out.

3.4.1. The OTA used as a resistor. For the implementation of full CMOS resistors the OTA in a unity feedback configuration has very often been used [17-19, 22]. The OTA-C lossy integrator is shown in Fig 3.7. For this circuit, the transfer function becomes

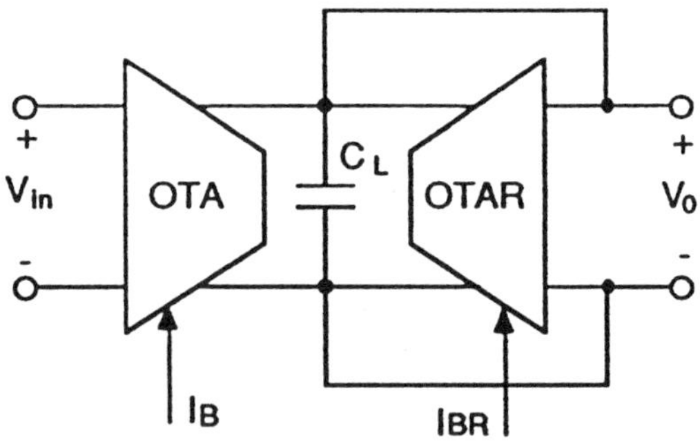

Fig. 3.7. The OTA-C lossy integrator. The resistor is implemented by the unity feedback OTA.

$$\frac{v_O}{v_{in}} = \frac{1}{\frac{g_{mr}}{g_m} + s\frac{C_L}{g_m}} = \frac{g_m}{g_{mr}}\frac{1}{1 + s\frac{C_L}{g_{mr}}} \tag{3.10b}$$

where the transconductances g_m and g_{mr} are associated with OTA and OTAR, respectively. This circuit presents several interesting properties. The dc gain of the lossy integrator depends of the transconductances ratio. As a result, its value depends of a ratio of dimensions and a ratio of the dc bias currents. Hence, it is almost insensitive to variations of both temperature and process parameters. Even more, for filters with low quality factors both transconductances, g_m and g_{mr}, are in the same order of magnitude. For these applications, OTAs with the same transconductance can be used [22]. Unfortunately, for filters with high quality factor large transconductance ratios are needed. In this case the precision of the transconductance ratio decreases. This is not a major limitation for these topologies because the active filters are much more sensitive to the imaginary part of the poles. These sensitivities are discussed in the next chapter. The pole of the lossy integrator is determined by g_{mr} / C_L.

Tuning range and distortion of the OTA based resistor. Typically, the integrator dc gain is tuned through the ratio of the transconductance g_{mr} to g_m. The pole is adjusted by g_{mr}. The OTA based resistor is controlled by its bias current as

$$R = \frac{1}{g_{mr}} = \frac{K_I}{\sqrt{2\,\beta_R\,I_{BR}}} \tag{3.11}$$

where K_I is a factor that depends of the OTA implementation, for instance see equation 2.15 for the LDOTA. Current I_{BR} is the bias current of the unity feedback connected OTAR. According to 3.11, the I_{BR} sensitivity of R is only 0.5. Hence, large resistor variations require even larger variations on the OTA bias current. Large variations on the OTA bias current have considerable effects on the dynamic range of the OTA. The distortion of the OTA based resistor can be computed using the same equations as for the OTA. In the previous sections it has been demonstrated that the most important distortion is HD3. From equation 2.17b of chapter 2, it can be noted that

$$HD3 \;\alpha\; \frac{1}{I_{BR}} \tag{3.12}$$

In this expression, the symbol α means "proportional to". The major disadvantages of these type of resistors can be explained from expressions 3.11 and 3.12. While the resistance depends of the square root of the bias current, the harmonic distortion is inversely proportional to the bias current. In the next chapter it will be shown that for narrow-band applications, the resistance have to be varied by more than 100 %. This is in order to compensate the excess phase errors of the OTAs. From relationship 3.12, if the bias current of the resistor decreases further then the harmonic distortion increases further as well. On the other hand, if the current increases the harmonic distortion described by 3.12 decreases. Nevertheless, for low supply voltages the common-mode range of the OTA is limited. As a result of this, a further increment on the bias current reduces drastically the common-mode range of the OTA. Hence, the dynamic range of the integrator is further reduced as well. For narrow-band applications, a different resistor implementations is a must.

3.4.2. Increasing the efficiency of OTA-C integrators. Regardless the limited tuning range of the OTA-C lossy integrator, this approach is the most commonly used for the design of low Q continuous-time filters [17-22]. In this section we will concentrate on the use of the OTA as a resistor. For the discussion, the topology of Fig. 2.8, chapter 2, is used as the reference.

Comments on the output-stage of the OTA. In the whole OTA, the noise is dominated by the transistors at the output-stage, as it has already been discussed in section 3.2. For high frequencies, however, large amounts of current are needed. If the bias current is increased, the transconductance of the OTA increases proportional to the square root of the current and the output impedance of the OTA decreases almost proportional to the bias current. As a result, the dc gain of the OTA reduces proportional to the square root of the bias current. The reduction of the OTA dc gain reduces the quality factor of the integrators and so reducing the quality factor of the filter as well. These effects among others are further discussed in the second section of the next chapter. Due to the reduction of the OTA dc gain, the use of cascode output-stages is unavoidable. In order to reduce the number of parasitic poles, the most commonly used output-stage is the folded cascode [7, 10, 11-13, 22, 23].

Typically in the folded cascode OTA the drain current of the output-stage is equal to the bias current of the input-stage. In Fig. 2.4, the current sourced by M_8 is equally split in the input-stage and M_5. It should be noted that the dc current flowing through M_5 flows to VSS through the transistor M_6. The current flowing through the transistors M_6 is purely dc current. Adding the contribution of all the transistors M_6, that dc current represents around 50 % of the OTA bias current. In other words, under these conditions the maximum efficiency of the folded cascode OTA is 50 %. For practical applications that efficiency is even lower.

For the OTA, the very high impedance output-stage is unavoidable, otherwise the quality factor of the integrators is further reduced. However, for the resistor the situation is different. A discussion of this question follows.

Avoiding the output-stage of the OTA. Several attempts to avoid the output-stage of the OTA have been made [18, 19]. In that technique, the output voltage is taken directly from the drain voltage of the transistors used in the OTA input-stage. For the case of a simple amplifier, while the input signal at the gate of the transistor increases the output signal at the drain of the same transistor decreases. The maximum difference between the input signal and the output voltage is determined by the threshold voltage of the transistor. As a result, the dynamic range of the OTA is further reduced in those topologies. Fortunately, this is not the case for the resistors. In order clarify the main idea, lets consider the circuit of Fig. 3.8.

This circuit is the input-stage of the LDOTA with the output connected to the input. Detailed analysis of the structure shows that the linear range of the "resistor" is limited by transistors M_3. This range is in the order of the threshold voltage of that transistor, typically in the order of 0.75 Volts if the bulk effects are neglected. If these effects are included, V_T increases to the range of 1 Volt. This range allows to use the "resistor" for differential voltages up to 4 V_{pp}, which is sufficient enough for most of the applications. The linearity is not degraded by the simplification of the structure. On the contrary, the linearity is slightly better than that obtained with the complete LDOTA. Simulations for a 10.7 MHz lossy integrator indicate that the linear range is not reduced by the simplification of the OTA indeed. The technique can also be applied to most of the OTAs proposed in the literature [18, 19, 22, 23].

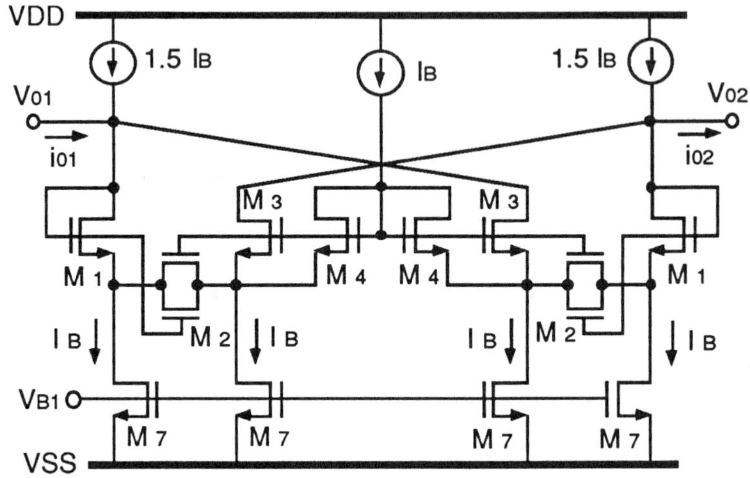

Fig. 3.8. Simplified OTA used as a fully-differential resistor.

In the simplified structure it can be argued that the resistor is sensitive to the output impedance of the transistors M_1 and M_3. Further, the finite output conductance of the current sources are also present in the resistor terminals. However, for filters wherein the quality factor is controlled by compensating the excess phase of the OTA, the finite output conductance's of the transistors may introduce a Q error in the order of 1 to 4 %. In other tuning schemes as those to be discussed in chapter 4, these effects can be further compensated.

3.4.3. MOSFET resistors. Another approach for the implementation of CMOS resistors uses transistors biased in linear region [20, 21, 24, 25], see Fig. 3.9. For a CMOS transistor biased in this region, its drain current is given by [20, 21]

$$i_d = \beta_1 \left(\left(v_g - V_{T0} + \gamma\sqrt{2\phi_F} - \frac{v_d + v_s}{2} \right)(v_d - v_s) - \frac{2}{3}\gamma\left(\left(v_d - v_b + 2\phi_F\right)^{1.5} - \left(v_s - v_b + 2\phi_F\right)^{1.5} \right) \right) \quad (3.13)$$

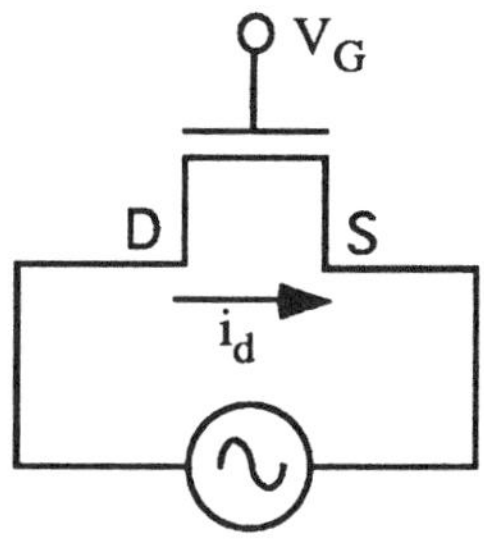

Fig. 3.9. The MOSFET resistor.

where v_g, v_b, v_d and v_s are the gate, bulk, drain and source voltages, respectively. V_{T0} is the threshold voltage, and ϕ_F the Fermi level. β_1 is determined by process and geometrical parameters. γ is the body factor, already defined in chapter 2. According to this equation, most of the drain current non-linearities are due to bulk, drain and source voltages. It has been shown that even with large drain and source voltage swings the second and third-harmonic distortions can be reduced to the order of 7 % and 1%, respectively [20, 21].

The MOS resistor has several very interesting properties that make it very attractive for linear applications. First of all, it is the simplest topology for the simulation of CMOS resistors and its noise factor is 1, the minimum achievable noise level. Its power consumption is low and, however, there are 3 terminals available. All of them can be used to control the drain-source current. In this section, the drain and the source are used as the resistor terminals and the tuning of the resistance is carried out by gate voltage. Nevertheless, other options can be exploited. The resistance of the MOS transistor biased in linear region is approximately given by

$$R_{ds} \cong \frac{1}{\beta_1 (V_G - V_T - V_{CM})} \tag{3.14}$$

where $V_{CM} = (V_D + V_S)/2$ is the common-mode voltage of the drain-source voltage. Typically, the resistance is controlled through the gate voltage. For low supply voltages, $V_G - V_{CM}$ is limited to around 2 Volts leading $V_G - V_{CM} - V_T \approx 1$ Volt. In this case, the resistance can be tuned in a range of ± 50 %. This tuning range is high enough for most of the low-Q applications. However, for high Q applications tuning range higher than ± 100 % are very often needed. Triode biased transistors have been successfully used in low-distortion transconductors as well [22, 23].

An important disadvantage of this topology is due to the effects of the common-mode voltage on the resistance. From relationship 3.14, it is clear that the resistance is sensitive to the common-mode offset voltage. Furthermore, the threshold voltage V_T is also sensitive to the common-mode voltage, V_{CM}. Hence the accuracy of the ratio of resistors is limited. This effect is present in all the circuits based on this type of transistors, such as transconductors. In these circuits, the dc offset voltage due to the common-mode feedback plays a very important role in the accuracy of the system. A partial solution for this problem is to control the resistor with a voltage $V_G = V_{G0} + V_{CM}$ [23]. The drawback of this approach is that the tunability range is further reduced and additional circuitry have to be used. Even more, additional parasitic poles are introduced to the system.

Distortion of the MOSFET. For the MOSFET based resistor, most of the non-linearities are even-order and so they can be reduced by subtraction using symmetrical structures. If the MOS transistor is connected to the output of a fully-balanced OTA, in the ideal case the even order non-linearities are canceled out [20,21]. From expression 3.13 and after several manipulations it can be shown that for fully-differential output signals, $v_d = -v_s$, the third-harmonic distortion of the MOS transistor biased in linear region is given by

$$\mathrm{HD3}(\gamma) \cong \left(\frac{\gamma}{384} \frac{1}{(V_G - V_{CM} - V_T)(2\phi_F - V_B + V_{CM})^{1.5}} \right) V_{ds}^{2} \tag{3.15}$$

From this equation and for supply voltages of ± 2.5 Volts, it is possible to achieve HD3 ≈ -60 dB even for v_{ds} voltages of the order of 2 V_{pp}. However, other second-order effects like mobility modulation increase the distortion. It should be recognized that the resistor terminals are the drain and source of the MOSFET. Hence, the resistor is sensitive to both lateral and vertical electric fields. If the bulk effects and the mobility modulation, due to the vertical and lateral electric fields, are taken into account then HD3 becomes

$$HD3 \cong HD3(\gamma) + \left(\frac{\theta'}{16\left(1 + \left(V_{GS} - V_T\right)\theta\right)^2} \left(\theta' + \frac{\eta}{V_G - V_{CM} - V_T} \right) \right) v_{ds}^2 \quad (3.15\text{ b})$$

where $\theta' = \theta + 2\,\mu / V_{max} L$ [4, 34]. θ is a fitting parameter to take in to account the reduction of the mobility due to the vertical electric field. For the 1.2 µm N-well process, θ is about 0.05 V^{-1} and 0.1 V^{-1} for the N and P transistors, respectively. In this case, the N transistors are less sensitive to the vertical electric field. The effects of the lateral electrical field are included in the term $2\,\mu / V_{max} L$; with μ, V_{max} and L the channel mobility, the saturation velocity and the length of the transistor, respectively. Because the larger mobility of the NMOS transistor, this transistor is more sensitive to this effect than the PMOS transistor. This factor is very often neglected, in most of the cases unjustly. For instance, for the NMOS transistor fabricated in the 1.2 µm process (see appendix A1.1 for process parameters) this factor is in the order of 1 / L (with L in microns). Hence, for L < 10 µm the effect of the lateral electric field dominates to the θ' effects. The parameter η has been already defined in chapter 2, relationship 2.13. Detailed definitions of these parameters can be found elsewhere [4, 6, 34]. Taking into account all these effects, the distortion of the MOS transistor can be as low as -58 dB for V_{ds} = 2 V_{pp}. Also, if the input signal contains a common-mode signal then the distortion increases rapidly. This highly common-mode signal sensitivity of the resistor distortion is evident in the expressions 3.15 and 3.15b.

3.4.4. The CMOS resistor based on saturated transistors. The CMOS resistors can be implemented using saturated transistors too. The main idea is to use the v_{gs}-i_{ds} characteristic of the transistor biased on saturation region [26, 27]. This relationship is of course not linear, but for small variations it can be approximated around the operation point as a linear function. In chapter 2 it has been shown that in the saturated differential pair the HD3 is inversely proportional to the bias current of the transistor and the transconductance of the transistor is proportional to the square root of the bias current. Hence, increasing further the bias current and using transistors connected in series the harmonic distortion components can be decreased to about -48 dB for input voltages up to 2 Vpp [27]. This approach, however, is limited in both linear range and tunability, mainly if low supply voltages are employed.

Linearization using current subtraction. Other techniques, using saturated transistors, take advantage of the quadratic law of the drain current. In saturation region, the transistor drain current can be approximated by the relationship

$$i_D \cong \frac{\beta}{2}\left(v_{GS} - V_T\right)^2 \tag{3.16}$$

The difference of the drain current of 2 matched transistors characterized by this expression becomes

$$i_{D1} - i_{D2} \cong \frac{\beta}{2}\left(v_{G1} - v_{G2} - v_{S1} + v_{S2}\right)\left(v_{G1} + v_{G2} - v_{S1} - v_{S2} - 2\,V_T\right) \tag{3.17}$$

If the input signal is applied to the gates, $v_{in} = v_{G1} - v_{G2}$, and if the source voltages are related to the gate voltages by a factor K, expression 3.17 reduces to a linear equation given by

$$i_{D1} - i_{D2} \cong \frac{\beta}{2}(1\text{-}K)\left[\,(1\text{-}K)\,V_{CM} - 2\,V_T\right]\,v_{in} \tag{3.18}$$

For fully-differential signals this expression becomes a linear function of the differential input signal. Several implementations for this expression have been already reported [26, 36]. However, those topologies are not discussed here. Nevertheless, some comments to this approach follows. First off all, the model of the transistor used for the deduction of 3.18 is very limited. The effects of the finite output impedance of the transistor and other second-order effects limits the validate of this expression. Secondly, the generation of the source voltages increases the complexity of the topology. Very often the poles and zeros introduced by the additional circuitry limits the frequency response of the structure. However, in 3.18, parameter K could be made zero. In this case the circuit becomes two totally independent and "matched" transistors with their source connected to a low impedance node, in most of the cases connected to the supply rails. Finally, expression 3.18 is based on the assumption of "matched transistors". Certainly, if the transistor mismatches are considered the harmonic distortion components increase rapidly. In general, the "linear" topologies based on the non-linearities cancellation through current subtraction of independent structures are very sensitive to transistor mismatches. This is because each structure generates large amounts of distortion, as is evident in the equation 3.16. When those currents are subtracted, a very small mismatch in the topologies results in considerable even order harmonic distortions. For the design of low-distortion resistors, other possible solutions are discussed in the next section.

3.5. Full CMOS resistors for continuous-time signal processors.

The first generation of IC continuous-time filters was based on OTAs and capacitors [17 - 19]. The OTA-C filters use unity feedback connected OTAs for the simulation of the resistors. Although the OTA is the mostly used building block, full CMOS resistors have recently been incorporated in continuous-time signal processors as well [11-13, 23, 34]. At this point, there are two interesting questions: Firstly, why CMOS resistors have been used? And the second important question is how to design low-distortion CMOS resistors and OTA-R-C filters? The aim of the next sections is to find out the answer to these questions.

3.5.1. Why linear CMOS resistors? The OTA based resistors are the most commonly used technique for the implementation of low-Q filters. The major reason is because for low Q applications the frequency response of the filters are not very sensitive to the OTA excess phase. Then, the quality factor of the filter can be controlled directly, using replicas of the OTA controlled with the automatic frequency control system. Unfortunately, this is not the case for narrow band applications, for which Q>10. In these applications, due to the effects of the parasitic poles and zeros and due to the process parameters tolerances, variations in the bandwidth of the filter even larger than 100 % can be expected. Because in the OTA based resistor the transconductance is controlled by the square root of the bias current, the compensation of these errors requires very large variations of its operation point. For low supply-voltages, e.g. ± 2.5 Volts, the common-mode range of the OTA is limited. If the current increases further, some of the transistors can reduce further the linear operation of the OTA. On the other hand, if the current decreases, the bias voltage of the input-stage decreases and then the THD increases further. Since the most important distortion is HD3, the variation of the THD is proportional to the square of the variation in the transconductance. Therefore, a reduction in the OTA transconductance of 100% implies an increment of 12 dB in the HD3. Further, the reduction of the bias current may reduce the frequency of the parasitic poles and parasitic zeros of the OTA. Hence, the phase response of the lossy integrator can be degraded.

The main idea behind the use of CMOS resistors is to have larger tunability of the resistor without sacrificing neither dynamic range nor the frequency response of the filter. Certainly, the linearity of the resistor has to be as good as the linearity of the OTA, otherwise the harmonic distortion components of the integrators will be dominated by the resistor non-linearities. It should be recognized that the introduction of the resistor in the filters have some additional implications. The resistor is an additional element used for the control of the quality factor of the filter. However, the tuning of the resistor requires a special controller. This controller increases the complexity of the system. Details of the nature of those controllers can be found in the next chapter.

The answer to how to design full CMOS resistors is given in the next sections. In the remaining of this chapter, several linearized resistors are described [30,32,34].

3.5.2. A low-distortion grounded resistor based on a triode biased transistor. Several OTA linearization techniques have been proposed in the literature but only a few for active resistors. The CMOS tunable resistors can also be simulated making use of transistors biased in linear region, as discussed in the section 3.4. In this section, a linear grounded resistor using a transistor biased in linear region is discussed.

A linear grounded resistor. For a CMOS transistor biased in the linear region, its drain current is approximately given by equation 3.13. From this equation it can be noted that the drain current of the MOSFET is a linear function of the gate voltage and the major non-linearities are due to the other transistor terminals. It is expected that the non-linearities decrease if the ac voltage in the drain, source and bulk terminals is minimized. This approach has been exploited in the design of very linear transconductors [28, 29]. If the gate terminal is connected to the output voltage while the source and bulk are connected to the supply voltage as shown in Fig. 3.10, a large-signal conductance can be defined as

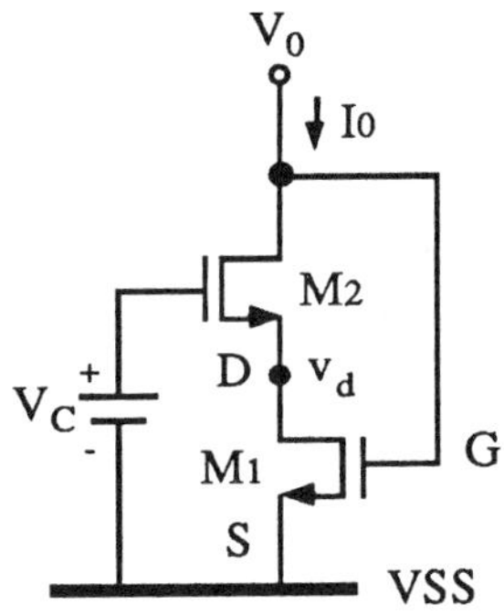

Fig.3.10. Linear grounded resistor.

$$g_0 = \frac{\partial i_d}{\partial v_g} = \beta_1\left(v_{DS1} + \frac{\partial v_{ds1}}{\partial v_g}\left(v_{GD1} - V_{T0} + \gamma\left(\sqrt{2\phi_F} - \sqrt{v_{DB1} + 2\phi_F}\right)\right)\right) \quad (3.19)$$

According to this expression, it is clear that the reduction of the harmonic distortion depends of the ability to maintain the ac-signal at node D as small as possible. For this circuit, if $\beta_2 >> \beta_1$, v_d is approximately given by $(g_{m1} / g_{m2})\, v_0$. Thus, increasing the transconductance of M_2 the harmonic distortion is further reduced. An important difference between this resistor and the MOSFET resistor, already discussed in the section 3.4, is that this resistor is little sensitive to the common-mode voltage of the input signal, $V_G = V_0$.

Tunability and distortion. In order to have a better idea of the tradeoffs involved in the resistor, detailed calculations have to be carried out. The linear range of the resistor is basically limited by the bias voltage V_C. It can be shown that the ac resistor voltage is limited to be smaller than the absolute value of V_G-V_D-V_{T1} (V_D=V_C+VSS-V_{GS2}). For VSS = -2.5 Volts and V_G=0, this range can be larger than 1 Volt. Neglecting the non-linearities, the resistance of the topology can be approximated by

$$R \cong \frac{1}{\beta_1 V_{DS1}} = \frac{1}{\beta_1 (V_C - V_{GS2})} \quad (3.20)$$

where V_{DS} is the drain-source voltage of M_1. V_C is the voltage to be used for the tuning of the resistor. The resistor can easily be tuned within a decade by adjusting V_{DS1} in the range of 0.1 to 1 Volt.

In expression 3.19, the most important non-linearities of the conductance are those due to the voltages v_{ds1} and v_{gd1}. These non-linearities can be reduced by reducing further the signal at the drain of M_1. For hand calculations only these non-linearities are considered. For the resistor, it can be demonstrated that the variation of the M_1 drain voltage is related to the gate voltage through the following equation, see appendix A3.1,

$$\frac{\partial v_{ds1}}{\partial v_g} \cong - \frac{g_{m1}}{g_{m2}(1+K)+g_{01}} \qquad (3.21)$$

where g_{m1} and g_{01} are the transconductance and the output conductance of M_1, respectively. Parameter g_{m2} is the transconductance of M_2 and the factor K is a result of the M_2 bulk effects. For hand calculations, these parameters can be approximated as

$$g_{m1} = \beta_1 V_{DS1} \qquad (3.22)$$

$$g_{01} = \beta_1 (V_{GS1} - V_{DS1} - V_{T1}) \qquad (3.23)$$

$$g_{m2} = \sqrt{2\,\beta_2\, I_{D2}} = \sqrt{2\,\beta_2\,\beta_1\, V_{DS1}(V_{GS1} - 0.5\, V_{DS1} - V_{T1})} \qquad (3.24)$$

$$K = \frac{\gamma}{2\sqrt{V_{DS1} + 2\,\phi_F}} \qquad (3.25)$$

Expression 3.21 is the classical gain voltage of an amplifier with transconductance g_{m1} and load impedance equal to $(1+K)g_{m2} + g_{01}$. It is interesting to note that the transconductance g_{m1} is proportional to V_{DS1} while g_{m2} is almost proportional to the square root of V_{DS1}. If V_{DS1} decreases, both parameters g_{m1} and g_{m2} decrease as well. In addition, the conductance g_{01} increases when V_{DS1} is decreased. The result is that the factor $(1+K)g_{m2} + g_{01}$ is little sensitive to V_{DS1}. In appendix A3.1 it is shown that the harmonic distortion components are proportional to the factor $(1+K)g_{m2} + g_{01}$. Hence, the harmonic distortion components of the resistor are lowly sensitive to V_{DS1}. Making a Taylor series expansion of relationship 3.19

and according to expressions 3.19-3.25, the second harmonic distortion of the resistor becomes

$$HD2 \cong \left(\frac{\sqrt{\dfrac{\beta_1}{2\beta_2}}}{(1+K)\sqrt{V_{DS1}\left(V_{GS1} - V_{T1} - \dfrac{V_{DS1}}{2}\right)} + \sqrt{\dfrac{\beta_1}{2\beta_2}}\left(V_{GD1} - V_{T1}\right)} \right) \frac{V_0}{2} \qquad (3.26)$$

The derivation of this expression and that for the third-harmonic distortion can be found in the appendix A3.1. If the effects of θ are neglected, HD2 can be further reduced by increasing the ratio of β_2 to β_1. According to equation 3.26, the variations on HD2 due to changes in V_{DS1} are slightly compensated, as previously discussed, by the several terms involved in the expression. The calculation for the third-harmonic distortion yields

$$HD3 \cong \left(\frac{2 + \sqrt{\dfrac{\beta_2}{2\beta_1}}\sqrt{\dfrac{V_{DS1}}{V_{GS1} - \dfrac{V_{DS1}}{2} - V_{T1}}}}{\left((1+K)\sqrt{\dfrac{2\beta_2}{\beta_1}}\sqrt{V_{DS1}\left(V_{GS1} - V_{T1} - \dfrac{V_{DS1}}{2}\right)} + \left(V_{GD1} - V_{T1}\right)\right)^2} \right) \frac{V_0^{\,2}}{8} \qquad (3.27)$$

Therefore, the distortion is slightly sensitive to the common-mode voltage of the incoming signal, V_G. In equations 3.26 and 3.27, it can be noted that the harmonic distortion components increase further for very low V_{DS1} bias voltages. In this case it is very difficult to have low distortion even if β_2 is further increased. This is because for very low V_{DS1} the factor $(1+K)\, g_{m2} + g_{01}$ is not compensated any more. For this situation, the transconductance g_{m2} reduces proportional to the square root of the bias current and the output conductance of M_1 is saturated to the value $\beta_1(V_{GS1} - V_{T1})$.

The distortion due to the mobility modulation is taken into account in the appendix A3.1, equations A3.6 and A3.7. For typical designs, this distortion is lower than the distortion due to the variations in V_{DS1}. In Fig. 3.11, results from HSpice simulations and hand calculations for the case $V_0 = 0.6$ Vpp are compared. For this case, the 1.2 μm process parameters have been used. The most relevant parameters are: $\theta = 0.1\ V^{-1}$, $\beta_1 / \beta_2 = 36$, $V_{GS1} = 2.5$ V, and $V_{T1} = 0.75$ V.

The second harmonic distortion of the grounded resistor is in the order of -40 dB. This harmonic distortion is further reduced, around -30 to -40 dB, when a floating resistor is considered. That topology will be discussed shortly. Although the very simple model used for hand calculations, the expressions for the HD3 predicts within 3 dB accuracy the behavior of the resistor. At very low V_{DS1} voltages, the predicted HD3 is higher than the HSpice result. At higher voltages the distortion obtained from HSpice increases faster than the harmonic distortion predicted by the expressions. The reason is because for $V_{DS1} > 1V$, the control voltage at the gate of M_2 is of the order or even higher than the common-mode voltage of V_0. Hence, the common-mode range of M_2 is further reduced. Obviously, this effect has not been included in the derivation of expressions 3.26 and 3.27.

Although the large value of θ, typically the third-order harmonic distortion is dominated by the V_{DS1} non-linearities. These non-linearities can be reduced by using additional feedback to the gate of M_2 [29]. This feedback increases the transconductance of M_2 and so it reduces further the variations on V_{DS1}. In that case HD3 is dominated by the mobility modulation effects. In general, the mobility degradation and the bulk effects are the last limitations for the implementation of ultra low-distortion voltage-to-current transducers.

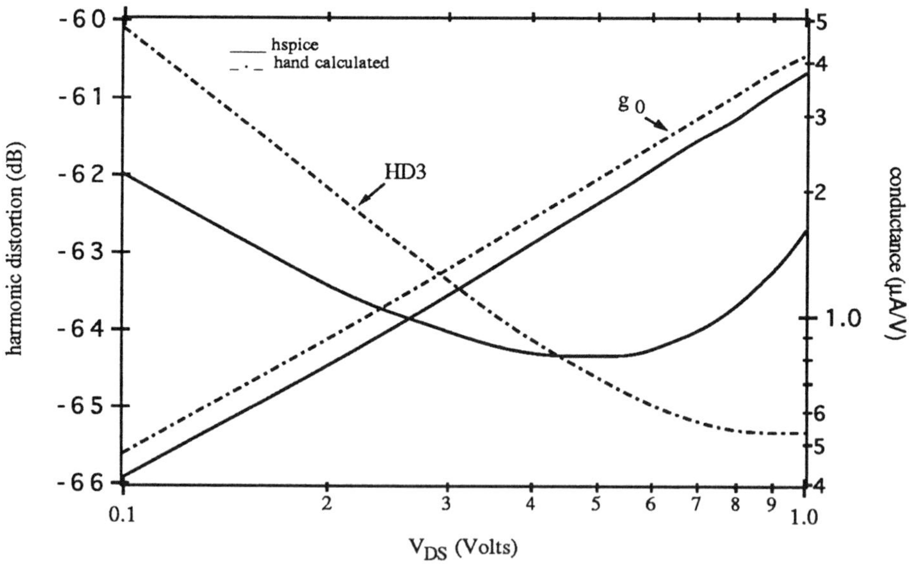

Fig. 3.11. HSpice and hand calculated results for the low-distortion grounded resistor.

Frequency response. The frequency response of the resistor is limited by the parasitic capacitors, mainly those capacitors associated with the node D (see Fig. 3.10). Namely, the total parasitic capacitor from D to ac ground, C_1, and the gate-drain parasitic capacitor, C_2, associated with M_1. The output admittance of the topology can be modeled as a frequency dependent conductance in parallel with a capacitor. The capacitor can be absorbed by the capacitors associated with the resistor terminals and their effect can be minimized by the automatic tuning system available in the OTA based filters. A small-signal analysis of the circuit shown in Fig 3.10 yields an ac output conductance given by

$$g_{0ac} \approx g_0 \left(1 - \frac{\omega^2 C_1 C_2}{g_0 \, g_{m2}} \right) \tag{3.28}$$

where g_{m2} is the transconductance of M_2 and ω is the frequency variable. g_0 is given by expression 3.19. The main limitation to these structures is the g_0 dependence of the resistor frequency response. The larger the resistance (smaller g_0) the smaller the -3 dB frequency. It must be remembered that in filtering applications the quality factor of the poles is very sensitive to the excess phase of the components. Hence, for these applications, it is advised to use this resistor at frequencies lower than f_{-3dB} / 10. For resistors in the range of hundred kΩ, the -3 dB frequency is in the range of 1 to 10 MHz.

3.5.3. Low-distortion floating resistor using triode biased transistors. This floating resistor is based on the grounded resistor discussed in the previous section. Comparing the expressions 3.26 and 3.27, it can be seen that the dominant distortion of the grounded resistor is HD2. The second harmonic distortion can be further reduced by current or voltage subtraction of symmetrical structures. This technique has been used in the LDOTA and in general in the fully-differential topologies. In this section a low-distortion floating topology, almost even distortion free, is presented.

The floating resistor is shown in Fig. 3.12 [12, 13]. In this circuit, four replicas of the grounded resistor, using PMOS transistors, are employed. The current of two of them is mirrored to the other resistor terminal making the effect of a real floating resistor and canceling the even-order harmonic distortions, except for mismatches.

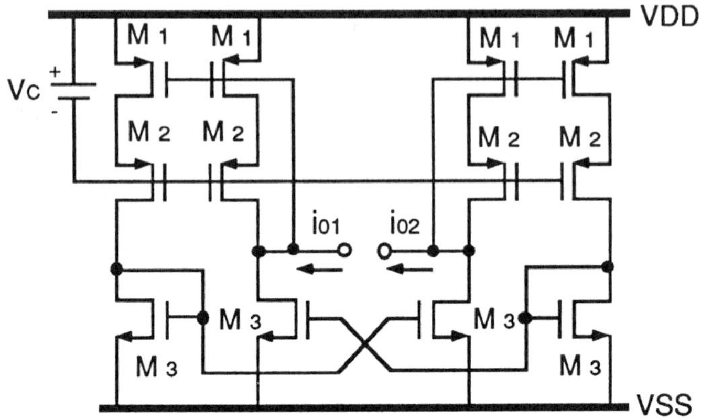

Fig. 3.12. Linear floating resistor based on linear region biased transistors.

Because the harmonic distortion components of the grounded resistors are diminished due to the effect of M_2, the mismatches between the symmetrical structures do not increases further the harmonic distortion components. For instance, if the HD2 of the grounded topology is around - 40 dB, then using the floating resistor this distortion is reduced by around - 34 dB if 2 % transistor mismatches are considered. Hence, the HD2 of the floating resistor is about - 74 dB. In this topology, for the even-order distortions there are two linearization procedures. The first one is due to the cascode transistor M_2 and the second one is due to the current subtraction of the replicas. For those reasons, the structure is little sensitive to the transistor mismatches.

The frequency response of the floating resistor is given by relationship 3.28. Typically the frequency of the poles due to the current mirrors are at higher frequencies and can be neglected. It should be recognized that the rejection to VDD noise is also higher, although the sources of the transistors biased in triode region are connected to the supply rail. This is because the drain of these transistors are also ac tied to VDD due to transistors M_2. For this reason, it is important that the impedance between the control voltage V_C and the supply rail has to be as low as possible.

Simulations. Some simulations for the floating resistor are briefly discussed here. A floating resistor of 520 kΩ has been designed and simulated using the 3 μm process, see appendix A1.1 for process parameters. The transistor dimensions (W/L) are 6 μm/18 μm and 120 μm/6 μm for M_1 and M_2, respectively. Transistors M_3 have the same dimensions as transistors M_2. The THD, including 2% transistor mismatches, is lower than -60 dB for voltages up to 2 Vpp. The -3 dB frequency is about 6 MHz. Using the minimum dimensions this frequency is around 10 MHz. As a conclusion, it can be said that this topology is useful for frequencies up to 1 MHz. A similar design has been used in the design of a fourth-order low-pass ladder filter, to be discussed in chapter 5.

3.5.4. Current source based grounded resistor. There are several techniques for the implementation of large resistors (= very small conductance's) most of them based on triode biased transistors. In this approach the resistance is proportional to the length of the transistor and hence its frequency response is limited, as it has been demonstrated in the previous circuits. Another technique that uses two OTAs connected in parallel but with the inputs interchanged, applying partial positive feedback, can also be used. In this technique, the equivalent transconductance is given by the difference between the transconductors connected in parallel. Hence, with large transconductors large resistors can be implemented. Obviously, the excess phase, the power consumption, the noise and the silicon area are increased. However, the major disadvantage of this approach is its high sensitivity to variations in the transconductance of the OTAs. The use of partial positive feedback has been used for long time in active RC filters [31]. The techniques using additional loops for the control of the Q and/or attenuators in the loops increase the excess phase, the noise level, the chip area and the power consumption [17]. Even more, due to the use of the attenuators, the dynamic range of the filter is reduced.

The saturated grounded resistor. The non-zero output conductance of the MOS transistor is the major limitation for the OTA and OPAMP dc gain. Typically, this conductance has been considered as an unwanted but unavoidable element in the transistor. In this section, this conductance is used for the control of the ac current flowing through the transistor. It is shown that using this approach the short comings of the previous techniques are avoided. Furthermore, this technique allows to adjust the resistance value in a very wide range. In addition, due to the absent parasitic poles its frequency response is extremely high. A simplified version of the saturated grounded resistor is shown in Fig. 3.13.

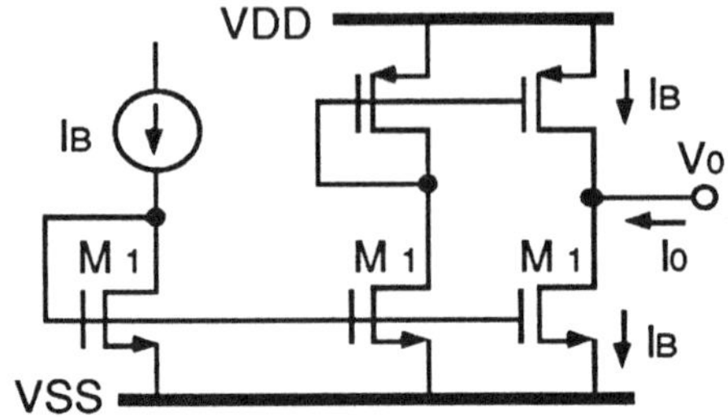

Fig.3.13. Low-distortion grounded resistor based on saturated transistors.

The output conductance of long channel transistors is proportional to the bias current and inversely proportional to the length of the transistor. For a typical CMOS 3 μm N-well process the Early voltage for NMOS transistors is in the order of 4.5 Volts / μm and twice for the PMOS transistors. The output admittance of the current source is the parallel connection of the output conductance of the N transistor and the conductance of the P transistor and their drain capacitors. These capacitors are not very important because they can be absorbed by the automatic tuning system, to be discussed in the next chapter. Hence, for a gate length transistors of 6 μm, the current source output conductance becomes

$$g_S = I_{BIAS}\left(\frac{1}{L_N V_{EN}} + \frac{1}{L_P V_{EP}}\right) \approx .055\; I_{BIAS}\; (A/V) \tag{3.29}$$

Some of the advantages of this technique can be explained from this equation. The implementation of resistors in the order of MΩ, conductance's in the range of a few μA/V, requires currents in the range of μA. Although the very low dc current used for the current sources, its frequency response is not degraded. This is due to the absence of parasitic poles. The mismatches in the current sources are not critical because the low current resulting can be easily absorbed by the high current in the output-stage of the OTA. The noise contributions of the transistors can be reduced by using small transconductances. A discussion of the linearity of this approach follows.

Detailed analysis of the transistor output conductance. In the most general case the transistor conductance is mainly due to both effects channel length modulation and drain-induced barrier lowering. The output conductance of the MOS transistor can be approximated by [4, 6, 33]

$$g_O = g_{OL} + \sigma\, g_m = I_B \frac{\partial}{\partial\, v_O}\left(\frac{\Delta L}{L}\right) + \sigma\, g_m \tag{3.30}$$

where I_B, L and v_O are the drain bias current, the effective length of the gate and the ac output voltage, respectively. Parameter g_m is the transconductance of the transistor and σ is a fitting parameter. Parameter σ is inversely proportional to the cubic of the transistor length, having a dominant effect in submicron transistors. For the used technologies and for conductance's higher than 2 µA/V, the conductance is dominated by the first term. In a simple case, $\Delta L / L$ can be approximated by [4]

$$\frac{\Delta L}{L} = \frac{X_D \sqrt{K\, V_X}}{L} \sqrt{1 + \frac{v_o}{V_X}} - \frac{K\, E_p\, X_D^2}{2\,L} \tag{3.31}$$

where $X_D = (2\, \varepsilon_s / qN_a)^{0.5}$. Parameters K and E_p, kappa and Early voltage (in V / µm), are Spice fitting parameters. Detailed definitions of these parameters can be found in the references [4,6,7,33]. The voltage V_X is given by

$$V_X = V_{DS} - V_{DSAT} + \frac{1}{K}\left(\frac{E_P X_D}{2}\right)^2 \tag{3.32}$$

with V_{DS} the drain-source bias voltage. For modern short channel CMOS processes and low supply voltages, V_X is dominated by the last term. As a result it is little sensitive to variations in V_{DS}. V_X is a strong function of the technology used and its value can not be accurately predicted. For a typical 1.2 µm process and with supply voltages of ± 2.5 Volts, its value is in the order of 2.5 to 4.5 Volts. From equations 3.30 and 3.31, the small-signal conductance becomes

$$g_O = \frac{I_B X_D \sqrt{K}}{2\,L \sqrt{V_X}} + \sigma \sqrt{2\,\beta_1 I_B} \tag{3.33}$$

Thus, the conductance can easily be tuned by the bias current I_B. Commonly this current is generated by the tuning circuits available in the continuous-time signal processor. From equations 3.30 to 3.33 and making a Taylor series expansion of the output conductance of the transistor, expression 3.30, the second and third-harmonic distortions are computed as

$$HD2 = \frac{1}{8\left(1 + \frac{\sigma g_m}{g_{0L}}\right)} \frac{V_0}{V_X} \tag{3.34}$$

$$HD3 = \frac{1}{32\left(1 + \frac{\sigma g_m}{g_{0L}}\right)} \left(\frac{V_0}{V_X}\right)^2 \tag{3.35}$$

where g_{0L} is the conductance due to the length modulation effect, the first term in equation 3.33. These equations predict lower harmonic distortions for lower bias current, larger resistor, because the term g_m/g_{0L} increases in this case. For $V_0 = V_X/10$, HD2 in the order of -40 dB and HD3 around - 70 dB are predicted. HD2 is in good agreement with the simulated results but HD3 is lower than simulations, - 65 dB. The reasons are: Firstly, the resistor is sensitive to the lateral electric field then the channel mobility is modulated by the drain voltage; secondly, in simulations the effect of the parasitic capacitors is also present. These effects have not been included in expressions 3.34. and 3.35.

3.5.5. The current source based floating resistor. For the floating resistor, techniques using saturated transistors have already been proposed [26, 27]. These resistors have higher frequency response than the CMOS resistors using triode biased transistors. In this section, a floating resistor based on the grounded resistor presented in the last section is discussed. Its resistance value is almost proportional to the bias current, leading to larger tunability, and the harmonic distortion is little sensitive to the resistance value. Furthermore, for fully-differential signals, the parasitic poles are at very high frequencies enabling its -3 dB frequency at frequencies in the range of several hundreds of MHz. As in the previous resistors, it is advised to use this topology at frequencies lower than f_{-3dB} / 10.

The topology. Regarding the characteristics of the current source based grounded resistor, the floating resistor shown if Fig. 3.14 is developed. The circuit looks like to the typical differential pair transconductor. However, the signal is not applied to the gates but to the drains. The gate terminals are ac grounded. It can be demonstrated by using circuit analysis that the differential output current is given by

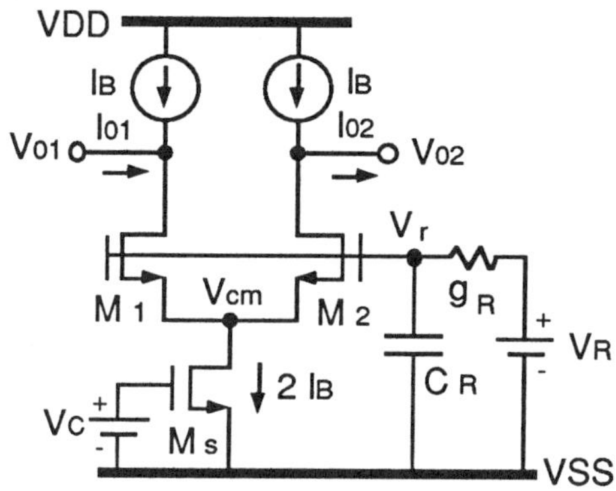

Fig.3.14. Linear floating resistor based on the output conductance of the saturated transistors.

$$
\begin{aligned}
i_{01} - i_{02} = & \frac{g_{01} + g_{02} + s\left(C_{gd1} + C_{gd2} + C_{db1} + C_{db2}\right)}{2}\left(v_{01} - v_{02}\right) \\
& + \frac{g_{01} - g_{02} + s\left(C_{gd1} - C_{gd2} + C_{db1} - C_{db2}\right)}{2}\left(v_{01} + v_{02}\right) \\
& + \left(g_{m1} - g_{m2} + s\left(C_{gd1} - C_{gd2}\right)\right) v_r \\
& + \left(g_{m2} - g_{m1} + g_{02} - g_{01}\right) v_{cm}
\end{aligned}
\tag{3.36}
$$

where g_{mi} and g_{0i} correspond to the transconductance and output conductance of the transistor M_i, respectively. C_{ijk} is the capacitance between the terminals i and j of transistor k. In this equation, s is the frequency variable $j\omega$. The first term is the ideal small-signal admittance of the topology and the other terms are a result of both transistor mismatches and finite impedance of the reference voltage. For fully-differential applications, $v_{01}=-v_{02}$, the second term vanishes. In practice, the amplitudes of v_{01} and v_{02} are not perfectly matched and thus a small common-mode component is always present. The real part of this component is almost not function of the frequency and thus the frequency response of the conductance is little affected. It can be demonstrated that v_r and v_{cm} are dominated by the common-mode output voltage. The differential signal component at these nodes is mainly due to mismatches in both gate-drain capacitors and drain-source conductance's of transistors M_1 and M_2. As the gate-drain capacitors of M_1 and M_2 are connected to C_R making a voltage divider, the high-frequency effect of these capacitors is further reduced because the capacitor C_R is much higher than the other capacitors present in the structure. So for fully-differential signals, the last three terms of equation 3.36 are quite small. As a result the frequency response of the conductance, the real part of 3.36, is very high. The drain capacitors can be considered as tolerances of the OTA load capacitor. It can be noted from Fig. 3.14 that each output of the resistor is connected to the VSS supply rail by a cascode impedance, determined by g_{mi}, g_{0i} and $2g_s$. So, the coupling to VSS is further reduced.

For fully-differential signals, the common node V_{cm} can be considered as ac ground. Thus, for the computation of the harmonic distortions, expressions 3.33 and 3.34 can be used at each terminal of the floating resistor. In the case of even-order distortions, they are further reduced when the output voltage is taken differentially. However, for this resistor the V_{DS} voltage is determined by the common-mode voltage of the signal and for the voltage at the common sources, V_{cm}. This voltage is controlled by the reference voltage V_R.

Simulations. For the simulations, a CMOS 1.2 μm N-well process has been used. Some process parameters are: $K = 0.5$, X_D=0.3 μm / $V^{1/2}$, E_p = 8.5 V / μm, σ = 0.01, V_X = 2.5 V and $g_m/g_{0L} \approx 50$. With these parameters, a 214 kΩ floating resistor has been designed and simulated. The dimensions, W / L, of the transistors M_1 and M_2 are 20 μm/1.5 μm. A 2 % mismatch in their lengths has been included. The dimensions of M_S are 20 μm/6 μm. The bias current is 50 μA and the reference voltage V_R is - 0.5 Volts. Similar results have been obtained for V_R=-1 Volt. The second, third and total harmonic distortions, at 10.7 MHz, versus differential input voltage are shown in Fig. 3.15.

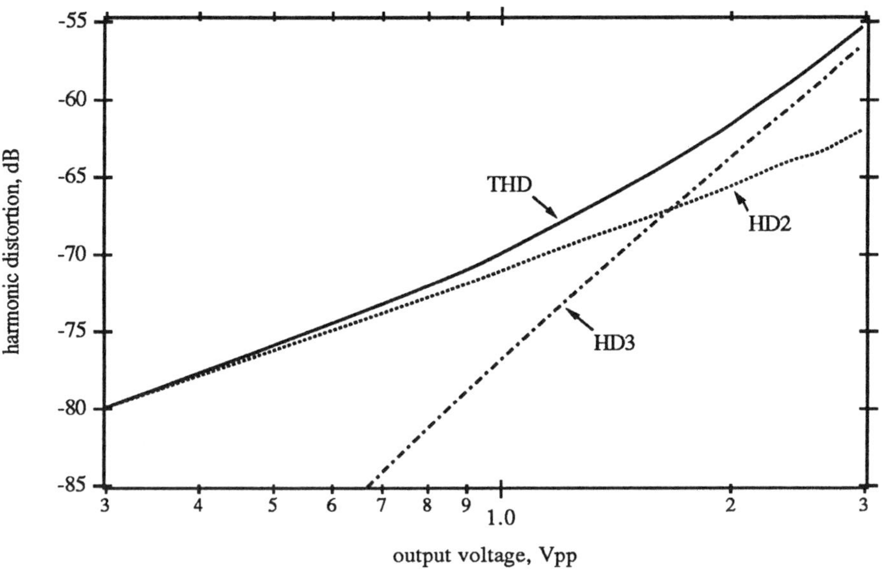

Fig.3.15. Harmonic distortions for the saturated floating resistor. 2% transistor mismatches have been included.

Summarizing the results, it can be said that the harmonic distortion is little sensitive to the resistor bias current. The conductance of the topology is almost proportional to the bias current and so it can be varied in a very width range. In fully-differential-mode, the total harmonic distortion is lower than -60 dB for input signals up to 2 Vpp. The variation of the resistance is lower than 1% for frequencies up to 100 MHz. This design has been used in a fourth-order bandpass filter [34]. The experimental results of that filter, discussed in chapter 5, have demonstrated the low distortion and wide tunability range of this technique.

3.6. Conclusions.

The design considerations for lossless integrators have been discussed. The effects of the non-linearities of the building blocks have been pointed out. The parasitic capacitors increase the load capacitor of the OTA. Therefore, those capacitors have to be incalculated in the poly-silicon capacitor. The effects of the non-linear junction capacitors have been considered as well. It has been demonstrated that for low-distortion integrators the poly-silicon capacitor has to be at least 4 times larger than the total junction capacitor associated with the integrator output. Also, a study of the integrator dynamic range has been carried out. A very simple equation, expression 3.6, allows to relate the filter design constraints with the practical lossless integrator requirements.

For the design of lossy integrators, several CMOS resistors have been discussed. The major requirements for the resistors are large tunability and good linearity. The accuracy of the absolute value of the resistor is not critical because the resistor is tuned by using an automatic tuning system. For filters, however, the accuracy of the ratio of resistors is much more important. In particular, the topologies discussed in section 3.5 allow to implement large resistances with harmonic distortion components of the order of -60 dB. In addition, the resistances are proportional to the bias current of the resistor and so they can be varied in a wide range, e.g. more than 1 decade. In particular, the current source based resistor achieves a very high-frequency response. This result is due to the reduced number of parasitic poles.

These topologies are very well suited for narrow-band applications wherein both small and large time constants are needed in the same filter. Typically, the small time constant is associated with the resonant frequencies of the filter and the large time constant is associated with the quality factor of the filter.

REFERENCES

[1] D. J. Allstot and W. C. Black, "Technological Considerations for Monolithic MOS Switched-Capacitor Filtering Systems," Proc. of the IEEE, Vol. 71, No 8, pp. 967-986, Aug. 1983.

[2] J. L. McCreary and P. R. Gray, "All-MOS Charge Redistribution Analog to Digital Conversion Techniques-Part I", IEEE J. of Solid-State Circuits, Vol. SC-10, pp. 371-379, Dec. 1975.

[3] P. R. Gray, D. A. Hodges and R. W. Brodersen, editors, *Analog MOS Integrated Circuits*, IEEE PRESS, New York, 1980.

[4] P. Antognetti and G. Massobrio, "*Semiconductor Device Modeling with SPICE*" Mc Graw-Hill Book Company, USA, 1988.

[5] P. Wambacq, G. Gielen, J. Vanthienen and W. Sansen, "A design Tool for Weakly Nonlinear Analog Integrated Circuits with Multiple Inputs," IEEE Proc. CICC-91, San Diego Cal., pp. 5.1.1 - 5.1.4, May 1991.

[6] Y. Tsividis, "*Operation and modeling of the MOS Transistor*", Mc Graw-Hill Book Company, Singapore, 1988.

[7] K. Laker and W. Sansen, "*Design of Analog Circuits and Systems*", Mc Graw-Hill Book Company, New York, to be published in 1993.

[8] W. Sansen, "Integrated Low-Noise Amplifiers in CMOS Technology," Nuclear Instrum. and Methods in Phys. Research, A252, pp. 427-433, 1987.

[9] Z. Y. Chang and W. Sansen, "Low-Noise Wide-Band Amplifiers in Bipolar and CMOS Technologies," Kluwer Academic Publishers, Massachusetts USA 1991.

[10] P. R. Gray and R. G. Meyer, " MOS Operational Amplifier Design-A Tutorial Overview," IEEE J. of Solid-State Circuits, Vol. SC-17, pp. 969-982, Dec. 1982.

[11] J. Silva-Martinez, M. Steyaert and W. Sansen, "A Large-Signal Very Low-Distortion Transconductor for High-Frequency Continuous-Time Applications," IEEE J. of Solid-State Circuits, Vol. SC-26, pp. 946-955, July 1991.

[12] J. Silva-Martinez, M. Steyaert and W. Sansen, "Design Techniques for High-Performance Full CMOS OTA-R-C Continuous-Time Filters," IEEE J. of Solid-State Circuits, Vol.27, No.7, pp. 993-1001, July 1992.

[13] M. Steyaert, J. Silva-Martinez and W. Sansen," High-Performance OTA-R-C Continuous-Time Filters with Full CMOS Low-Distortion Floating Resistors," IEEE Proc. ESSCIRC-91, Milano Italy, pp. 5-8, Sept. 1991.

[14] F. E. Girling and E. F. Good, "Active Filters 7 and 8: The Two Integrator Loop," Wireless World, Vol. 76, pp. 117-119, March 1970, and pp. 133-139, April 1970.

[15] F. E. Girling and E. F. Good, "Active Filters 12 and 13: The Leapfrog or Active Ladder Filters," Wireless World, Vol. 76, pp. 354, May 1970, and pp. 445-450, July 1970.

[16] A. S. Sedra and P.O Brackett, "*Filter Theory and Design: Active and Passive*", Matrix Publishers, Inc., Champaign Illinois, 1978.

[17] R. L. Geiger and E. Sanchez-Sinencio, "Active Filter Design Using Operational Transconductance Amplifiers: A Tutorial," IEEE Circuits and Devices Magazine, Vol. 1, pp. 20 - 32, March 1985.

[18] F. Krummenacher and N. Joehl," A 4 MHz CMOS Continuous-Time Filter with On-Chip Automatic Tuning," IEEE J. of Solid-State Circuits, Vol. SC-23, pp. 750-757, June 1988.

[19] H. Khorramabadi and P. R. Gray," High Frequency CMOS Continuous-Time Filters," IEEE J. of Solid-State Circuits, Vol. SC-19, pp. 939-948, Dec. 1984.

[20] Y. Tsividis and P. Antognetti, editors, "*Design of MOS VLSI Circuits for Telecommunications*", Prentice-Hall, Inc., Englewood Clifs, New Jersey, 1985.

[21] M. Banu and Y. Tsividis, "Fully integrated active RC filters in MOS technology," IEEE J. of Solid-State Circuits, Vol. SC-18, pp. 644-651, Dec. 1983.

[22] V. Gopinathan and Y. P. Tsividis," A 5V 7th-Order Elliptic Analog Filter for Digital Video Applications," IEEE Int. Solid-State Circuits Conference, San Francisco, pp. 208-209, Feb. 1990.

[23] Y. T. Wang, F. Lu and A. A. Abidi,"A 12.5 MHz CMOS Continuous Time Bandpass Filter," IEEE Int. Solid-State Circuits Conference, New York, pp. 198-199; Feb. 1989.

[24] K. Nagaraj, "New CMOS floating voltage-controlled resistor," IEE Electronics Letters, Vol. 22, pp. 667-668, June, 1986.

[25] M. Banu and Y. Tsividis, "Floating voltage-controlled resistors in CMOS technology," IEE Electronics Letters, Vol. 18, pp. 678-679, July 1982.

[26] S. P. Singh, J. V. Hanson, and J. Vlach, "A new floating resistor for CMOS technology," IEEE Trans. on Circuits and Systems, Vol. CAS-36, pp. 1217-1220, Sept. 1989.

[27] Z. Wang, "Novel electronically-controlled floating resistors using MOS transistors operating in saturation", IEE Electronics Letters, Vol. 27, pp. 188-189, January 1991.

[28] J. L. Pennock, "CMOS triode transconductor for continuous-time active integrated filters," IEE *Electronics Letters*, Vol. 21, PP. 817-818, Aug. 1985.

[29] B. Stefanelli and A. Kaiser,"CMOS Triode Transconductor with High Dynamic Range," IEE Electronics Letters, Vol. 26, pp. 880-881, June 1990.

[30] J. Silva-Martinez, M. Steyaert and W. Sansen, "Very linear CMOS floating resistor," IEE Electronics Letters, Vol. 26, pp. 1610-1611, Sept. 1990.

[31] R. Tarmy and M. S. Ghausi, " Very High-Q Insensitive Active RC Networks," IEEE Trans. on Circuit Theory, Vol. CT-17, pp. 358 - 366, Aug. 1970.

[32] M. Steyaert, J. Silva-Martinez and W. Sansen," High Frequency Saturated CMOS Floating Resistor for Fully-Differential Analogue Signal Processors," IEE Electronics Letters, Vol. 27, pp. 1609-1611, Aug. 1991.

[33] S. Liu and L. W. Nagel, "Small-Signal MOSFET Models for Analog Circuit Design," IEEE J. of Solid-State Circuits, Vol. SC-17, pp. 983-998, Dec. 1982.

[34] M. Steyaert and J. Silva-Martinez, "A 10.7 MHz CMOS OTA-R-C Bandpass Filter with 68 dB Dynamic Range and On-chip Automatic Tuning," IEEE Int. Solid-State Circuits Conference, San Francisco, 66-67, Feb. 1992.

[35] K. Bult and H. Wallinga," A Class of Analog CMOS Circuits Based on the Square-Law Characteristic of an MOS transistor in Saturation," IEEE J. of Solid-State Circuits, Vol. SC-22, pp. 357-365, June 1987.

Further readings.

[36] P. M. Van Peteghem and R. L. Rice, "New CMOS Resistor Implementation for Linear IC Applications," IEE Electronics letters, Vol. 24, pp. 288-290, March 1988.

[37] M. Banu and Y. P. Tsividis, "Detailed Analysis of Non-idealities in MOS Fully Integrated Active RC Filters based on Balanced Networks;" IEE Proceedings, Vol. 131, G, pp. 190-196, Oct. 1984..

[38] P. E. Allen and D. R. Holberg, "*CMOS Analog Circuit Design*", Holt, Rinehart and Winston, New York: NY, 1987.

[39] R. Gregorian and G. C. Temes, "*Analog MOS Integrated Circuits for Signal Processing*", John Wiley & Sons, New York:NY, 1986.

Appendix A3.1

Harmonic distortion for the grounded resistor based on a triode biased transistor.

In this appendix the harmonic distortion components for the grounded resistor of section 3.5.2 are computed. It is assumed that M_2 can be characterized by the ideal quadratic law. Then, for M_2, both mobility modulation and finite output impedance are neglected. For M_1, however, the current is characterized by expression 3.13. From this expression, the large-signal transconductance of M_1 is equal to the v_g gradient of the drain current, which is given by

$$\frac{\partial i_{d1}}{\partial v_g} = \beta_1\left(v_{DS1} - \frac{\partial v_{ds1}}{\partial v_g}\left(v_{GD1} - V_{TO} + \gamma\left(\sqrt{2\,\phi_F} - \sqrt{v_{DB1} + 2\,\phi_F}\right)\right)\right) \tag{A3.1}$$

The most important non-linearity of the transconductance is due to the term v_{gd}. For hand calculations, only this non-linearity is considered. From this relationship, the second and third gradients of the drain current can be easily calculated as

$$\frac{\partial^2 i_{d1}}{\partial v_g^{\,2}} \cong 2\,\beta_1 \frac{\partial v_{DS1}}{\partial v_g} \tag{A3.2}$$

and

$$\frac{\partial^3 i_{d1}}{\partial v_g^{\,3}} \cong 3\,\beta_1 \frac{\partial^2 v_{DS1}}{\partial v_g^{\,2}} \tag{A3.3}$$

From these expressions, it is clear that the distortions, proportional to expressions A3.2 and A3.3, depend of the ac-signal at the drain voltage of M_1. From fundamental circuit analysis, it can be demonstrated that the variation of the M_1 drain voltage is related to the input voltage through the following equation:

$$\frac{\partial v_{ds1}}{\partial v_g} \cong -\frac{g_{m1}}{g_{m2}\,(1+K) + g_{01}} \tag{A3.4}$$

where g_{m1} and g_{01} are the transconductance and the output conductance of M_1, respectively. g_{m2} is the transconductance of M_2 and the factor K is due to the bulk effects of M_2. For hand calculations, these parameters can be approximated as

$$g_{m1} = \beta_1 V_{DS}$$

$$g_{01} = \beta_1 (V_{GS} - V_{DS} - V_{T0})$$

$$g_{m2} = \sqrt{2\beta_2 I_{D2}} = \sqrt{2\beta_2\beta_1 V_{DS}\left(V_{GS} - 0.5\, V_{DS} - V_{T0}\right)}$$

$$K = \frac{\gamma}{2\sqrt{V_{DS} + 2\phi_F}}$$

Expression A3.4 is the classical gain voltage of the circuit. In the same form, the second derivative of v_{ds1} can be approximated by

$$\frac{\partial^2 v_{ds1}}{\partial v_g^{\,2}} \cong \left|\frac{\partial v_{ds1}}{\partial v_g}\right| \left(\frac{2\beta_1 + \sqrt{\dfrac{V_{DS1}}{V_{GS1} - 0.5\, V_{DS1} - V_{T1}}}}{g_{m2}(1+K) + g_{01}} \right) \qquad \text{(A3.5)}$$

In addition to the non-linearities already considered, there are those non-linearities due to the effects of the lateral and vertical electrical fields. Because the small-signal at the drain of M_1, the lateral field effects are quite small and can be neglected. From equations A3.1, A3.2, A3.4 and taking into account the effects of the vertical electrical field, the second harmonic distortion becomes

$$\mathrm{HD2} \cong \left(\frac{\sqrt{\dfrac{\beta_1}{2\beta_2}}\left(1 + \theta\,(V_{GS1} - V_{T1})\right)}{(1 + K)\sqrt{V_{DS1}\left(V_{GS1} - V_{T1} - \dfrac{V_{DS1}}{2}\right)} + \sqrt{\dfrac{\beta_1}{2\beta_2}}\left(V_{GD1} - V_{T1}\right)} \right)\frac{V_0}{2} + \qquad \text{(A3.6)}$$

$$+ \frac{\theta}{1 + (V_{GS1} - V_{T1})\theta}\,\frac{V_0}{2}$$

and the third-harmonic distortion yields

$$HD3 \cong \left(\frac{2 + \sqrt{\dfrac{\beta_2}{2\,\beta_1}}\sqrt{\dfrac{V_{DS1}}{V_{GS1} - \dfrac{V_{DS1}}{2} - V_{T1}}}}{\left((1+K)\sqrt{\dfrac{2\,\beta_2}{\beta_1}}\sqrt{V_{DS1}\left(V_{GS1} - V_{T1} - \dfrac{V_{DS1}}{2}\right)} + \left(V_{GD1} - V_{T1}\right)\right)^2} \right) \frac{v_0^{\ 2}}{8} + \qquad \text{(A3.7)}$$

$$+ \frac{\theta\, HD2\, v_0}{2\left(1 + |V_{GS1} - V_{T1}|\,\theta\right)}$$

Although of the limited models used, these expressions are in good agreement with results from Spice.

CHAPTER

4 THE ON-CHIP AUTOMATIC TUNING OF CONTINUOUS-TIME FILTERS

4.1. Introduction.

In this chapter we go up to the system level. The OTAs, the resistors and the capacitors are considered as the building blocks. They are represented by a simple macro-model but including the parameters that reproduces the main characteristics of the original block. This is a behavioral modeling of the building blocks [1-6].

Most of the fully-integrated continuous-time filters use OTAs and capacitors as the main elements. Due to the lack of virtual grounds and low impedance nodes, the continuous-time filters are sensitive to the parasitic capacitors. The time constants of the filter depend of C/g_m ratios. These time constants, unlike to the switched-capacitor circuits, are not very well controlled in CMOS technologies because they are affected by the process parameter tolerances. These tolerances can be more than 30% if temperature variations, parasitic capacitors, bulk effects and other effects are included. In addition to these, the active filters are sensitive to the OTA finite parameters, e.g. parasitic poles and zeros and finite dc gain [4-6,13-14].

In a first approximation, the magnitude of g_m/C determines the frequency of the poles (ω_o). Usually, the variations of this ratio are corrected using a Phase Locked Loop (PLL) [8-19, 21]. A replica of a two integrator loop is included in the PLL in order to track g_m/C to an external clock frequency. Typically, it is not necessary to tune all the loops of the main filter because ω_o is little sensitive to the OTA excess phase shift. Furthermore, the matching of time constants of the same order of magnitude can be very well controlled in CMOS technologies. Unfortunately, the Quality factor (Q), or the BandWidth (BW), of the filter is very sensitive to the OTA excess phase shift and to its finite dc gain. These errors can be partially corrected using magnitude detection and/or phase detection techniques [9, 14-16]. For better matching, it is preferred that the operating frequency of both tuning loops be equal to the resonant frequency of the main filter. This last fact increases the tuning signals feedthrough, mainly in high-frequency applications. Reported results for filters tuned by the PLL based tuning system have shown that the tuning feedthrough noise is even higher than the noise level of the filter [13, 14, 16, 21]. This shortcoming is partially solved if the control loops are operated in the stop-band frequency range of the filter. The drawback of this approach is that the tracking between the controller and the main filter is limited , e.g. 1 - 4 %.

In a different approach, it is proposed to tune ω_O and Q using a different frequency range than this of the filter resonant frequencies [22, 23]. The frequency tuning technique is based on a charge comparison principle. The processing of the charge is carried out by using well known switched-capacitor techniques [25]. Furthermore, because phase comparison is avoided, the tuning range of this technique is very large. In this technique, ω_O is tracked to an external clock reference times a ratio of two dc current sources. The ratio of the dc current sources allows to scale down (or up) the operating frequency of the control loop. The accuracy of this system is comparable to the accuracy of the PLL based techniques. For the BW tuning, the pulse response of the filter is exploited. In biquadratic filters, for $Q>0.5$, the envelope of its pulse response is an exponential function of the filter bandwidth. Tracking this envelope with a pole frequency equal to half of the bandwidth of the biquadratic filter, the filter bandwidth can be tuned.

In this chapter, the effects of the OTA finite parameters on a biquadratic filter are studied. It is shown that the frequency of the poles is little sensitive to the OTA non-ideal parameters. It is also shown that the filter bandwidth is sensitive to those parameters, mainly for narrow band filter applications. As a result, for high-frequency and narrow band applications the Q tuning of the filter is unavoidable. All these effects are further discussed in section 4.2. The phase locked loop based tuning techniques are considered in section 4.3. Both techniques, voltage controlled filter based PLL and voltage controlled oscillator based PLL, are studied and their advantages and limitations are pointed out. The charge based technique for the automatic frequency tuning of continuous-time filters is considered in section 4.4. The design considerations and the limitations of the technique are further discussed as well. For the tuning of the BW, the transient response of the filter is exploited. The technique is described in section 4.5. The limitations of the technique and some possible solutions are addressed as well. At the end of the chapter, some conclusions are drawn.

4.2. The second-order filters.

The CMOS OTA-C filters can be designed using either the ladder technique or the biquadratic sections approach [1 - 6]. In order to be able to compare the ladder filters with the biquadratic based filters, a simple second-order filter is considered. However, most of the results apply to higher-order filters as well.

4.2.1. The biquadratic loop. The topology shown in Fig. 4.1 results from the simulation of a second-order RLC low-pass filter [1-5]. In this topology, the transfer function at the outputs V_{01} and V_{02} correspond to a non-zero dc gain band-pass filter and to a low-pass filter, respectively. If the conductance g_1 is zero, then the low-pass ladder filter reduces to the typical variable state filter: the Biquad [4 - 6]. The higher-order band-pass ladder filters are not based on this topology using $g_1=0$. However, for higher-order band-pass filters based on biquads, this topology with $g_1 = 0$ is the main building block. Hence, in this text, this topology (with $g_1=0$) will be called the biquad. In this case the output V_{01} and V_{02} correspond to a zero dc gain band-pass and a low-pass transfer functions, respectively.

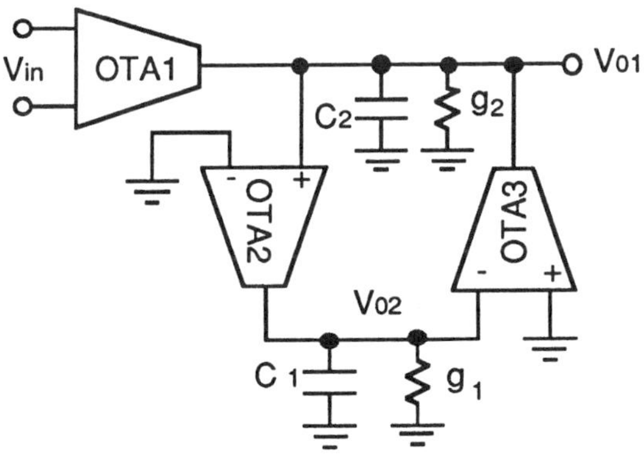

Fig. 4.1. Typical second-order filter.

Using circuit analysis and considering ideal OTAs, the transfer functions of the topology shown in Fig. 4.1 are calculated as

$$A_{V01} = \frac{\frac{g_{m1}}{C_2}\left(\frac{g_1}{C_1} + s\right)}{s^2 + \frac{\omega_0}{Q}s + \omega_0^2} \tag{4.1}$$

and

$$A_{V02} = \frac{\frac{g_{m1}g_{m2}}{C_1C_2}}{s^2 + \frac{\omega_0}{Q}s + \omega_0^2} \tag{4.2}$$

where ω_0 and Q are the poles frequency and the quality factor of the poles, respectively. These parameters, among of others, are given in table 4.1. In this table, $BW = \omega_0/Q$ is the filter bandwidth.

parameter	ladder	biquad $(g_1 = 0)$
ω_0	$\sqrt{\frac{g_{m2}g_{m3} + g_1g_2}{C_1C_2}}$	$\sqrt{\frac{g_{m2}g_{m3}}{C_1C_2}}$
BW	$\frac{g_1}{C_1} + \frac{g_2}{C_2}$	$\frac{g_2}{C_2}$
Q	$\frac{\sqrt{g_{m2}g_{m3} + g_1g_2}}{\sqrt{\frac{C_2}{C_1}}\,g_1 + \sqrt{\frac{C_1}{C_2}}\,g_2}$	$\frac{\sqrt{g_{m2}g_{m3}}}{g_2}\sqrt{\frac{C_2}{C_1}}$
A_{V01} @ dc	$\frac{g_{m1}g_1}{g_{m2}g_{m3} + g_1g_2}$	0
A_{V02} @ dc	$\frac{g_{m1}g_{m2}}{g_{m2}g_{m3} + g_1g_2}$	$\frac{g_{m1}}{g_{m3}}$
A_{V01} @ $\omega = \omega_0$	$\frac{g_{m1}}{C_2}\frac{\sqrt{\left(\frac{g_1}{C_1}\right)^2 + \omega_0^2}}{\omega_0\left(\frac{g_1}{C_1} + \frac{g_2}{C_2}\right)}$	$\frac{g_{m1}}{g_2}$
A_{V02} @ $\omega = \omega_0$	$\frac{\frac{g_{m1}g_{m2}}{C_1C_2}}{\omega_0\left(\frac{g_1}{C_1} + \frac{g_2}{C_2}\right)}$	$\frac{g_{m1}}{g_2}\sqrt{\frac{g_{m2}}{g_{m3}}}\sqrt{\frac{C_2}{C_1}}$

Table 4.1. Ladder against biquad: the most important filter parameters.

Technological considerations. Before to discuss further these parameters, it is desirable to consider some technological constraints. It has already been demonstrated by several authors that in CMOS technologies the capacitor ratios are more precise than the transistor transconductance ratios [27 - 29]. For a unity capacitor ratio, a 0.1 % accuracy can be achieved. The accuracy of the transistor transconductance ratios is of the order of 0.25% if large bias voltages are used and if the source-bulk voltages of the transistors are equal to zero. In addition, the precision of the transistor transconductance ratios is proportional to both gate area and gate-source bias voltage [27-29]. Thus, the higher the bias voltage and the gate area of the transistors the higher the precision of the transconductance ratio is. In the case of applications with reduced supply voltages and moderate transconductance ratios, less than 10, accuracy's of about 0.5 % are commonly found. In the case of complex circuits, as the OTA or the LDOTA, and larger transconductance ratios the precision is even lower, e.g. 1 to 2 %. With these comments in mind, the filter parameters shown in table 4.1 are discussed now.

The frequency of the poles. In the ladder filter, ω_0 is function of both the loop transconductances, g_{m2} and g_{m3}, and the resistors. In the biquad, ω_0 is not affected by the resistor. Because the frequency of the poles involve several parameters, the tuning of the ladder filter is more complex and less precise than the tuning of the biquad. For the same reason, the poles frequency of the ladder filter is less sensitive to the particular variations of each of these parameters. Hence, the frequency of the poles is less sensitive to the transconductance tolerances for the ladder filter than for the biquad, as could be expected [1, 2, 4, 5].

The bandwidth of the filter. In both cases, BW is controlled by the resistors. It should be noted that the ladder filter is controlled by both conductance's, g_1 and g_2. As a result, for the same BW, the ladder filter requires smaller conductance's, larger resistors, than the biquad. This is a considerable disadvantage of ladder filters, mainly for narrow-band filters wherein large resistors are needed. In the previous chapter it was shown that for large resistances the frequency response of several CMOS resistors is limited. Hence for high-frequency and narrow-band applications the biquad is more attractive.

Quality factor. The Q of the filter is given by the ratio of ω_0 and BW. In the biquad, the conductance g_2 is used to tune the Q of the filter. Therefore, the Q tuning of the biquad is independent of the poles frequency. This is not the case for the ladder filter wherein any adjustment of the resistors affects the frequency of the poles. This fact complicates the automatic tuning of the ladder filters.

dc gain. For the ladder filter, at dc the ratio of Av_{01} to Av_{02} is equal to g_1/g_{m1}. If $g_{m1}=g_1$, then both gains are equal, leading to an optimum filter dynamic range at dc. For the biquad the case is exactly the opposite. The V_{01} dc gain is zero and the V_{02} dc gain is given by g_{m1}/g_{m3}. This is a major disadvantage of the biquad: at dc and low frequencies its dynamic range is very bad. For low-pass filter applications, this fact reduces drastically the dynamic range of the biquad.

Gain at the resonant frequency. In the ladder filter, the gain voltage at both outputs is slightly different. Nevertheless, it is always possible to scale the filter such that the gain at both outputs become very close to each other. If so, the dynamic range of the ladder filter is optimum for the whole low-pass frequency range, from dc up to ω_0. For the biquad, both gains are function of the input transconductor g_{m1}, a transconductance ratio and a capacitor ratio. If $g_{m2}=g_{m3}$ and $C_1=C_2$, both gains are equal, resulting in an optimum filter dynamic range at the resonant frequency. In this case the peak gain is determined by g_{m1}/g_2. Note that g_{m1} is independent of the other filter parameters, namely, the pole frequency and the quality factor and the bandwidth. Hence, the peak gain of the biquad can be tuned by using g_{m1}.

It can be concluded that the former technique, ladder, gives better pass-band sensitivity to components variations. However, the second technique allows to control almost independently the resonant frequency, the quality factors and the peak gain of the filter. For low-pass filters, the dynamic range of the ladder filter is optimum. This is not the case for the biquad wherein huge low-frequency attenuation in the band-pass output occurs. For band-pass applications, the dynamic range of both filters is similar. Even more, it is not totally clear that for high-Q applications the ladder filters are less sensitive to component tolerances than the biquads [4]. However, the higher programmability of the biquad makes it very attractive for tunable continuous-time band-pass filter applications. Moreover, for narrow band applications, the biquad requires lower resistance values than the ladder filter.

4.2.2. Effects of the OTA finite dc gain and parasitic poles and zeros. The importance of the OTA excess phase, due to the finite dc gain and parasitic poles and zeros always present in the OTA, in the design of filters has been pointed out by several authors [4-6, 13-14, 17]. In this section, the effects of the finite

OTA parameters on the center frequency and on the bandwidth of the filter are considered.

Effects on the resonant frequency of the filter. If only the dominant pole, the dominant zero and the finite output conductance of the OTAs are considered, the resonant frequency of the biquadratic loop can be approximated as [22]

$$\omega_0 \cong \sqrt{\frac{g_{m2}\,g_{m3}+g_1\,g_2}{C_1 C_2}}\left[1+\frac{\dfrac{1}{A_{V2}}\dfrac{g_2}{g_{m3}}+\dfrac{1}{A_{V3}}\dfrac{g_1}{g_{m2}}}{2\left(1+\dfrac{g_1\,g_2}{g_{m2}\,g_{m3}}\right)}-\frac{g_{m2}\,g_{m3}}{2\,C_1 C_2}\left(\frac{1}{\omega_z}-\frac{1}{\omega_p}\right)^2\right] \qquad (4.3)$$

where g_{mi} and $A_{vi}=g_{mi}/g_{oi}$ are the transconductance and the dc gain of the ith OTA. For the gain A_{v3}, g_{03} have to be replaced by $g_{01}+g_{03}$ due to the effect of the output conductance of OTA1 (see Fig. 4.1). If N OTAs are connected to the same node, the total load conductance on that node is given by the addition of the whole N load conductance's. If the OTAs are identical then the OTA3 dc gain becomes A_{V3}/N, with A_{V3} the dc gain of the unloaded OTA3. In 4.3, ω_z and ω_p are the frequencies of the dominant zero and dominant pole of the OTAs. It is considered $\omega_z = \omega_{z2} = \omega_{z3}$ and $\omega_p = \omega_{p2} = \omega_{p3}$. From 4.3, after several algebraical manipulations, it can be obtained that the normalized deviation of the poles frequency due to the finite dc gain of the OTAs can be expressed as

$$\frac{\Delta\omega_0}{\omega_0}(A_v) \cong \frac{1}{2\,Q\,A_{V3}} \qquad (4.4)$$

According to this expression, the ω_0 deviation is higher for low-Q factors. For low Q applications, $Q < 10$, the error depends of the finite dc gain of the OTA3 (=OTA2). If low-frequency deviations are required, large OTA dc gains have to be used. This constraint makes the use of cascode output stages very desirable, unavoidable for high-performance applications. For this reason, and others that will be apparent shortly, in this text only folded cascode OTAs are used.

From 4.3, the effect of the OTA parasitic poles and zeros on the resonant frequency of the loop satisfies

$$\frac{\Delta\omega_0}{\omega_0}(\omega_{eq}) \leq -\frac{1}{2}\left(\frac{\omega_0}{\omega_{eq}}\right)^2 \tag{4.5}$$

where $\omega_{eq} = 1/(\omega_z^{-1} - \omega_p^{-1})$. Note that ω_{eq} may be positive or negative or infinite. This equivalent frequency depends of the pole and zero frequencies. According to this equation, ω_0 is little sensitive to the parasitic poles and zeros of the OTA2 (OTA3). Even if the equivalent parasitic frequency ω_{eq} is only 10 times higher than ω_0 the deviation on ω_0 is about 0.5 %. For better ω_0 accuracy it is desirable to place the OTA parasitic poles and zeros at frequencies higher than 20 times ω_0. Hence, the effect of ω_{eq} is a deviation on ω_0 lower than 0.15 %.

Effects on the quality factor of the filter. The quality factor of the filter is also affected by these OTA parameters. If the OTA finite parameters are included in Fig. 4.1 and using fundamental circuit analysis, the quality factor of the filter can be obtained as follows [22]

$$\frac{1}{Q} \cong \frac{\dfrac{g_1}{C_1} + \dfrac{g_2}{C_2}}{\omega_0} + \frac{\dfrac{g_{01} + g_{03}}{C_2} + \dfrac{g_{02}}{C_1}}{\omega_0} + 2\frac{\omega_0}{\omega_{eq}} \tag{4.6}$$

Typically, the quality factor of the filter is controlled with g_1 and/or g_2. From this equation it can be obtained that even in absence of parasitic zeros and poles the maximum Q obtainable with this loop is limited by the finite dc gain of the OTAs. Combining equations 4.3 and 4.6, it can be shown that the maximum Q is limited to around $0.3A_{V2}$-$0.5A_{V2}$. Because for high-frequency applications the OTA dc gain is further reduced, this constraint let us conclude the unavoidable use of cascode output stages. In the case that C_1 is equal to C_2, expression 4.6 reduces to

$$Q \cong \frac{Q_0}{1 + \dfrac{g_{01} + g_{02} + g_{03}}{g_1 + g_2} + 2Q_0\dfrac{\omega_0}{\omega_{eq}}} \tag{4.6b}$$

with Q_0 the ideal Q factor, given by the expressions of the third row of table 4.1. The importance of the OTA excess phase is evident in 4.6b. Even if the equivalent frequency of the parasitic poles and zeros, ω_{eq}, is in the order of 100 ω_0, which is quite difficult to achieve for high frequencies, its effect is quite severe. In that case and if the frequency of the zero is dominant as is the case in the LDOTA, the quality factor is limited to values lower than 50. From 4.6b, the normalized Q deviation of the biquad due to ω_{eq} can be computed as

$$\frac{\Delta Q}{Q}\left(\omega_{eq}\right) \cong -\frac{2Q\dfrac{\omega_0}{\omega_{eq}}}{1+2Q\dfrac{\omega_0}{\omega_{eq}}} \tag{4.7}$$

From this expression, the effects of the equivalent frequency are noticeable. If $\omega_{eq} < 0$ the parasitic pole is dominant, then the Q deviation is positive if $2Q\omega_0/\omega_{eq} < 1$. This effect results in an enhancement of the filter Q factor. The result is a peaking at the resonant frequencies of the filter. If $\omega_{eq} > 0$, then the behavior is dominated by the parasitic zero. In this case the Q deviation is negative, resulting in some attenuation on the filter gain at the resonant frequency. In 4.7 it can be noted that the Q deviation is extremely large when $2\,Q\,\omega_0/\omega_{eq}$ is close to -1. Some normalized deviations for realistic cases are given in table 4.2.

Q	$\frac{\omega_0}{\omega_{eq}}$	$\frac{\Delta Q}{Q}$ (%)
10	- 0.01	25
20	- 0.01	67
20	- 0.02	400

Table 4.2. Q deviations due to the filter Q and OTA excess phase shift.

In this table a negative ω_{eq} is assumed. From this table it can be seen that for high Q applications, extremely large errors are introduced by the excess phase shift of the OTA. Even more, these errors are for a single biquad. In higher order filters these effects are reflected just in the pass-band of the filter. Therefore, both peak gain and ripple of the filter are further affected. It is evident that for narrow band filters, the tuning of the filter Q is unavoidable.

4.2.3. The unavoidable tuning procedure for the filter. The deviations of the earlier active filters were corrected adjusting some resistors or variable capacitors [24]. This was possible because these designs were made with discrete components. In integrated circuits, a similar approach is very often used for the trimming of analog to digital and digital to analog converters. In this case a laser trimming procedure for the tuning of a resistor string is employed. However, the trimming procedure increases further both cost of the device and complexity. In integrated circuits, the continuous-time filters require a trimming procedure too.

Regarding the analysis carried out in the previous sections, in continuous-time filters the frequency of the poles is determined by a transconductance to capacitor ratio, e.g. g_m/C. A detailed consideration of this ratio leads to

$$\frac{g_m}{C} = K \frac{\sqrt{2\, C_{oxt} \mu_n \frac{W_t}{L_t} I_B}}{C_{oxc} W_c L_c} \tag{4.8}$$

where the subscripts t and c are for parameters associated with the OTA transistors and the capacitor, respectively. The channel mobility μ_n is a function of the substrate doping and the temperature. K is a topology dependent factor (see chapter 2 section 2.3). A more complete model should include the body factor effects and the parasitic capacitors as well. These effects produce tolerances on the transconductance of the order of 5 to 10 %, see sections 2.3 and 3.2. If process parameters tolerances, bulk effects, parasitic capacitors and temperature variations are accounted the frequency of the poles has variations by more than 30 % from the designed frequency. These variations change from device to device and from batch to batch. Under these conditions, the only possible application of these filters is restricted to a very large tolerance antialiasing filters. In order to extend the applications of continuous-time filters, these deviations have to be corrected. For integrated circuits, it is very desirable to use a self correcting system, otherwise these errors have to be corrected manually. Using the automatic tuning systems, the precision of the continuous time filters can be increased to the range of 1 to 5 %. Unfortunately, it is quite difficult to increase this accuracy. The main limitations are due to the mismatches in the filter building blocks and also due to the tracking errors between the tuning system and the continuous-time filters. However, the mismatches between the "identical" building blocks used in both tuning system and filter increase the tuning errors.

In equation 4.8, the only user parameter that can be used for the correction of the tolerances is the OTA bias current. Another possibility is to use capacitor banks instead of single capacitors. However, the discrete nature of the banks should require larger silicon area. In addition, the precision of this approach is limited by the resolution of the capacitor banks [20]. Other techniques tune the filter directly [18, 19]. In this case, both input and output of the filter are sampled. The magnitude and the phase of both signals are extracted and compared with certain references. From those comparisons the self correcting signals are generated. A major drawback of this technique is that for very small signals, or in absence of signal, the tuning of the filter can not be guaranteed. Hence, when the signal is larger again, the filter and the tuning system need some time for stable operation. This disadvantage makes this approach impractical.

In recently published designs, the correction of the process parameters tolerances and the temperature variations have been carried out by a master and slave systems. The self correcting system adjusts the bias current of the master and slave OTAs such that ω_0 is tracked to a stable and very well controlled reference [7-16, 21-23,31]. The accurate and stable reference is commonly a clock frequency. A discussion of these tuning techniques follows.

4.3. Automatic tuning techniques based on Phase Locked Loops.

In continuous-time filters, a clock frequency is the most commonly used reference. The major reasons for that are its highest precision, its high stability and because the clock is, typically, available in the system for the digital design. As the reference is a clock frequency, the typical schemes for the automatic tuning of continuous-time filters are based on the Phase Locked Loop (PLL) principle. The PLL based tuning techniques are discussed in this section.

4.3.1. PLL using a band-pass filter. The schematic of the PLL based controller using a band-pass filter is shown in Fig. 4.2. In this scheme, the voltage controlled filter (master filter) is a replica of a section of the main filter (slave filter). For an ideal biquad, at the resonant frequency the peak gain and the phase shift of the band-pass filter are g_{m1}/g_2 and zero or 0 degrees, respectively. The parameters g_{m1}

and g_2 are the transconductance of the first OTA and the conductance of the lossy integrator, respectively (see Fig. 4.1 and table 4.1). In the automatic tuning system, there are two loops; the Automatic Frequency Controller (AFC) and the Automatic Gain Controller (AGC). In the AFC, the master filter is phase locked to the reference signal. The phase of the reference signal is compared with the phase response of the biquad by the phase comparator. The phase error is filtered out by the low-pass filter and fed back to adjust the transconductance of the OTAs involved in the master filter. Ideally, the loop is locked when the phase of the band-pass filter output signal is equal to the phase of the reference signal. As a result, the phase shift of the voltage controlled band-pass filter at ω_0 is tuned to zero degrees. This is ideal shift phase at ω_0 of a second-order band-pass filter.

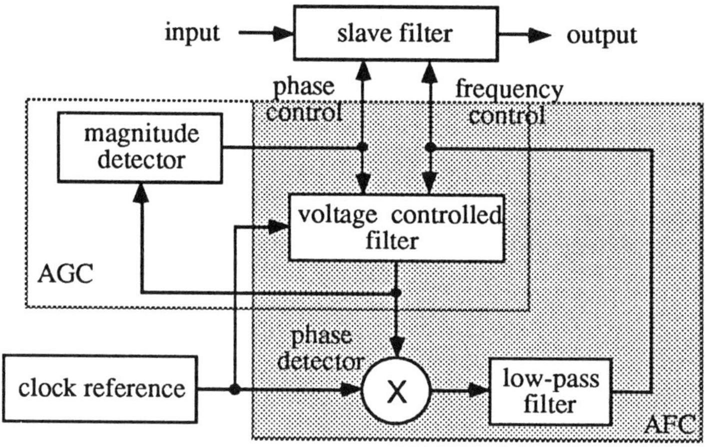

Fig. 4.2 Automatic tuning system based on a voltage controlled filter.

Adjusting the bandwidth of the filter. In the ideal case, the AFC loop guarantees that the phase shift of the band-pass filter at the resonant frequency is zero degrees. In absence of either parasitic poles or parasitic zeros, this means that the frequency of the poles ω_0 is equal to the clock frequency. The frequency ω_0 is determined by the magnitude of the poles indeed. However, the quality factor of the filter is determined by the real part of the poles. Hence, Q is not tuned by the AFC loop. For low Q applications, the biquads are little sensitive to the Q errors. For these applications, only the AFC loop is used. If higher precision is required, the automatic gain control has to be used for the tuning of the slave filter as well [14-16, 21].

The peak gain of the band-pass filter is controlled by the transconductance of the first OTA and the bandwidth of the filter (see table 4.1). Hence, sensing the amplitude of the filter response and comparing this with the amplitude of the reference signal the peak gain of the band-pass filter can be adjusted. However, this adjustment guarantees that the ratio of g_{m1}/C_1 to the filter bandwidth is unity. Because g_{m1}/C_1 can be tuned through the frequency controller, the filter bandwidth is indirectly tuned. This scheme does not guarantee 100 % tuning of the filter bandwidth. Nevertheless, its accuracy is sufficient enough for most of the applications. A simplified block diagram of this tuning scheme is shown in Fig. 4.3.

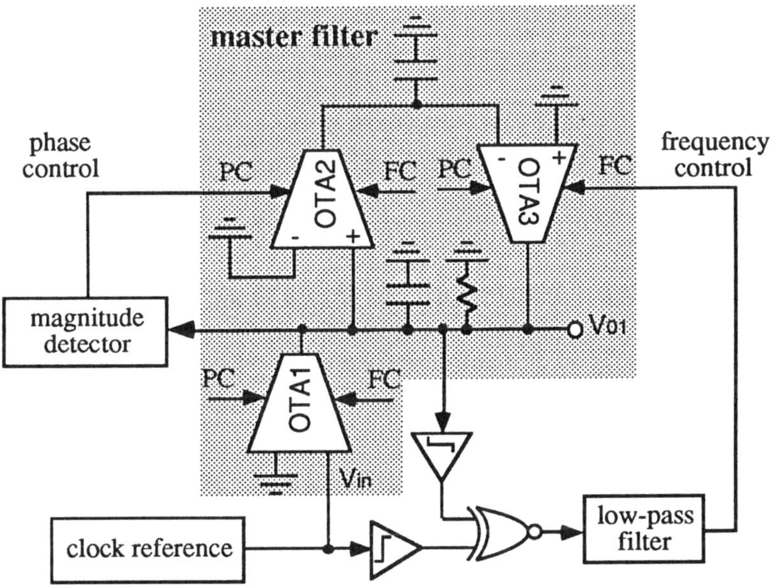

Fig. 4.3. Self tuning system based on a voltage controlled band-pass filter.

In this tuning scheme, the master filter is implemented as a typical biquad band-pass filter. The phase detector is realized by the voltage comparators and the NEXOR gate. In Fig. 4.3, the frequency control signal FC adjusts the bias current of the OTAs. The tuning of the biquad bandwidth is carried out through the adjustment of the integrators excess phase shift. This procedure adjusts ω_{eq} such that BW is tuned to the desired value (see equation 4.6). For this purpose, other networks have to be added to the OTA [16]. A lead phase network can be implemented as shown in Fig. 4.4.

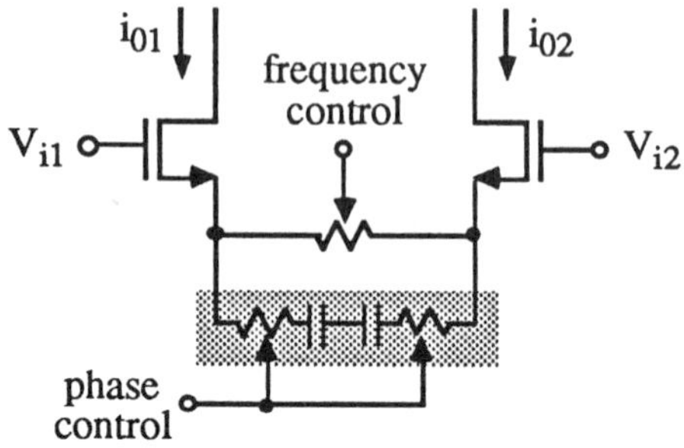

Fig. 4.4. A source degenerated OTA input stage with a lead phase network.

The active resistors can be implemented with either one of the topologies discussed in chapter 3. In general, these networks introduce either excess lead phase or excess lag phase in the frequency response of the OTA. Adjusting the phase of these networks, ω_{eq} in relationships 4.6 an 4.6b, the magnitude response of the biquad is tuned. Because the peak gain of the biquad is correlated to its bandwidth, see table 4.1, then the BW is indirectly tuned. These techniques will be further discussed shortly.

Limitations of the technique. The band-pass based PLL uses phase comparison for the correction of the filter frequency response. In addition, magnitude detection is employed for the correction of the filter shape. The main task of the AFC system is to adjust the phase shift of the band-pass filter. However, it is well known that the group delay of the band-pass filters is maximum at the center frequency of the filter [24]. The group delay of the filter is defined as the frequency derivative of the filter phase response. At the resonant frequency the group delay of the biquad is equal to 2/BW. Also, it is well known that the lock range of the PLLs is limited to around ± 45 degrees [7]. In fact, it was already demonstrated that the lock range of this system is limited to the range $f_0 \pm BW/2$ [13]. Because the process tolerances can be as larger as ± 30 %, this technique is useful for applications where $Q < 2$.

In addition to the limited frequency range of the previously discussed technique, there are other drawbacks. In this technique, the phase response of the filter is compared with the phase of the incoming signal. From this comparison the correcting error voltage is generated. In the ideal case this error voltage is zero. In the real situation, this error voltage is inversely proportional to the loop gain. The effect is the same as the error voltage at the input stage of the OPAMP based circuits. As a result, the finite loop gain of the PLL introduces tuning errors. For low Q filters, a loop gain of 150 introduces a tuning error of the order of 0.1 % [13]. For higher Qs, the error increases almost proportional to the Q value. In order to increases the dc gain of the loop, a high dc gain amplifier has to be employed. However, the introduction of the high dc gain amplifier and low dc offset voltage is not always an easy design task. The dc offset voltage of the dc amplifier decreases the precision of this approach. Another effect present in the structure is due to the distortions of the clock reference. The harmonic components of the reference signal are modulated by the band-pass output signal and so intermodulation harmonic distortion is generated. Some of those distortion components are dc components (see section 2.2 chapter 2). This dc noise produces errors on the dc output of the phase detector. It has to be remembered that the dc voltage is the control voltage for the slave filter. This shortcoming can be further reduced if a sinusoidal reference signal instead of the clock signal is used. The drawback is that an additional filter for the reference signal is needed.

The major disadvantages of this technique are because both the precision of the system and the loop controlling signal depend of the phase comparison. As a result, most of the dc offset voltages present in the loop affect the precision of the structure. A less sensitive offset voltage technique is the voltage controlled oscillator based controller.

4.3.2. PLL Using a Voltage Controlled Oscillator. In this approach, the PLL is based on a Voltage Controlled Oscillator (VCO). A block diagram of this system is depicted in Fig. 4.5. In this technique, the important parameter is the frequency comparison instead of the phase comparison. Both the reference frequency and the oscillating frequency are compared by the phase detector. The error voltage is filtered out by the low-pass filter and fed back to control the transconductance of the OTAs. Under locked condition, the oscillating frequency of the VCO is tracked to the clock frequency. If the biquad of Fig. 4.1 is employed, with g_1 and g_2 equal to zero, this condition implies that

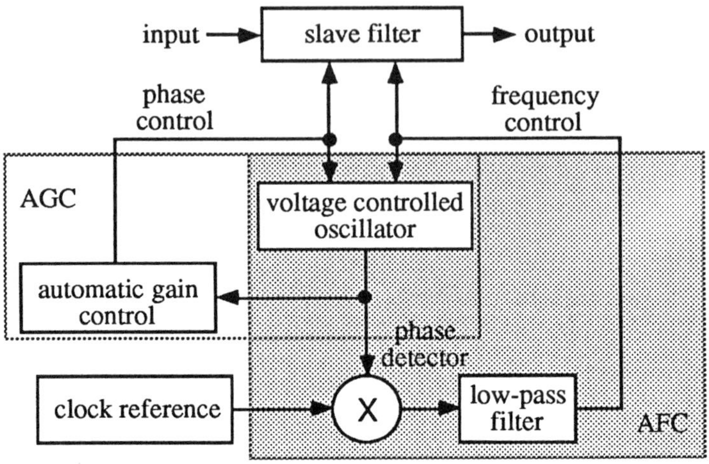

Fig. 4.5. Automatic Tuning system based on a voltage controlled oscillator.

$$\omega_0 \cong 2\pi f_{clock} \cong \sqrt{\frac{g_{m2}\, g_{m3}}{C_1 C_2}} \tag{4.9}$$

or

$$2\pi f_{clock} \cong \frac{g_{m2}}{C_1}\sqrt{\frac{g_{m3}}{g_{m2}}\frac{C_1}{C_2}} \tag{4.9b}$$

with f_{clock} the frequency of the clock reference. According to 4.9, the pole frequency g_{m2}/C_1 is tracked to $2\pi f_{clock}$. The frequency of the other poles can be tuned by capacitor and/or transconductance ratios. It has to be noted that this technique is little sensitive to the offset voltages of the loop. The only effect of the loop offsets is to decrease the dynamic range of the PLL but they do not degrade the accuracy of the system. Typically the total dc offset voltage of the loop is of the order of a few millivolts. So, its effect can be neglected. This property makes this approach less sensitive to the intermodulation effects and also lowly sensitive to the phase detector non-linearity. However, this is true if and only if the output voltage of the VCO is well controlled. Typically, an automatic gain control has to be included in the system. The result is that using this approach the requirements for the phase detector are further relaxed. Hence, a simple EXOR gate can be used as the phase detector. However, the reference signal can be a simple clock signal.

The tuning of the bandwidth. In Fig. 4.5, the automatic gain control adjusts the amplitude of the VCO output. The AGC is used for the control of the loop excess phase [16,21]. For this purpose a lead-lag network is included in the OTA. The poles and zeros of this network are adjusted to have stable amplitude oscillations. In this case, the phase shift of the VCO loop at $\omega=\omega_0$ is 180 degrees. This is the excess phase shift introduced by two ideal lossless integrators connected in a loop. In this approach, the bandwidth of the filter has to be adjusted by capacitor and/or transconductance ratios. For low Q filters, the quality factor of the filter is tuned by transconductance ratios.

Design considerations. Regarding the characteristics of both techniques, the VCO technique is less sensitive to the non-idealities of the controlling loop. However, care has to be taken the design of the loop. For instance, the linear range of the OTA is limited. Because the oscillating frequency of the VCO is related to the transconductance of the OTA, the VCO output voltage has to be such that the OTAs are always operating in their linear region. The non-linearity due to the automatic gain control can be critical because they affect the fundamental VCO frequency. The effect of these non-linearity are similar to the effects produced by the OTA non-linearity. However, these effects can be reduced by a carefully design of the AGC.

Some general comments are:

1) The PLL configuration guarantees that the oscillating frequency of the VCO is equal to the fundamental frequency of the reference clock signal. Also, due to the automatic gain control, the gain and the phase shift of the loop at this frequency are unity and 180 degrees, respectively.

2) The product of the integrators gain at the resonant frequency is unity. This does not necessarily imply that the gain of each one is unity. This is the case if and only if both integrators are exactly equal. Typically, this is not the case in practical circuits because the additional impedances associated with one integrator of the loop. These impedances are due to both phase comparator and automatic gain control. If additional positive and negative resistors are used for the automatic gain control, these impedances unbalance the integrators as well.

3) The same comments apply to the phase of the loop. Namely, the total phase shift of the loop is 180 degrees but this does not necessarily mean that the phase shift of each integrator is 90 degrees.

4) In order to have good matching with the slave filter, the effects of the additional load impedances present in both VCO and slave filter should also be

considered. Hence, the matching of the major components like OTA transconductance and poly-silicon capacitors is not the only important design consideration. For a better matching the impedances due to the OTA input and output stages and the number of OTAs connected to the outputs of the filter have to be taken into account when dimensioning the poly-silicon capacitors associated with the master filter.

5) The last resolution for the accuracy of the master-slave technique is due to the unavoidable mismatches. The mismatches are present independently of the tuning technique employed. These effects among others will be discussed shortly.

4.3.3. General comments on the PLL based techniques. In this section the major properties of the automatic tuning systems based on PLLs are briefly discussed. Most of these remarks apply to all the master-slave based tuning techniques. For instance, the mismatches between the master and the slave filters are present whether the PLL tuning technique is used or not. In the next sections different approaches for the automatic tuning of continuous-time filters are discussed including the mismatches.

The problem of the mismatches. Both short distance and long distance mismatches introduce tuning errors in the main and slave filter. The short-distance mismatches are those due to the edge effects of the transistor gate. Those edge effects are due to the lateral diffusions, birds beak effect, etc. This type of mismatch produces variations on the beta of the transistor. Also, the fixed gate oxide charge produces mismatches in the threshold voltage of the transistors [25]. The long distance mismatches are due to the gate oxide thickness gradients and also due to the substrate concentration gradients. Both factors affect the threshold voltage and the beta of the transistors. The bulk effects increase the transistor mismatches as well. Using special layout techniques and placing the master system as close as possible to the slave filter these errors can be reduced. Statistical analysis and experimental results for transistors have shown that these mismatches can be reduced to 0.5 % using both large bias voltages, $V_{GS}-V_T \approx 1$ V, and large gate areas, W/L ≈ 20μm/20μm, [27-29]. In continuous-time filters, however, these approaches can not always be used because the limited supply voltages. In the previous chapters it was shown that for low supply voltages the maximum $V_{GS}-V_T$ voltage used in the OTA is about 0.6 V. For typical OTAs using a minimum gate length of 6 μm, it is expected to have OTA transconductance mismatches in the order of 1%. However, the mismatches increase if shorter transistors are employed. According to the published results, the transistor mismatches are inversely proportional to the gate area [27-29]. Hence, using large gate area transistors the mismatches can be reduced. The drawback of this approach is that the parasitic capacitors increase as well and hence

reducing the frequency response of the system.

On the realization of the bandwidth controller. In the first generation of the CMOS continuous-time filters, only the frequency tuning system was used. However, in the most recent publications some phase compensating networks have been included as well [14-16, 21]. The compensating networks are based on a first-order pole-zero system. The transconductance of the OTA input stage using this type of networks is generally described by the following expression

$$g_m(s) = g_m \frac{1 + \dfrac{s}{\omega_z}}{1 + \dfrac{s}{\omega_p}} \tag{4.10}$$

with g_m, ω_z and ω_p the OTA low-frequency transconductance, the frequency of the equivalent zero and the frequency of the equivalent pole, respectively. Typically, the dc gain of the compensating network is unity. The phase response corresponds to the phase equalizer behavior. At low and very high frequencies, the phase of the network is zero degrees. However, at medium frequencies the phase behavior depends on the pole and zero frequencies. If the frequency of the pole is lower than the frequency of the zero, the circuit works as a first-order low-pass filter. Hence, it presents excess lag phase. In the other case, $\omega_p > \omega_z$, the circuit introduces excess lead phase. These networks are included in the voltage-to-current converter leading in the most general case to the integrator gain given by

$$A_{V_{int}}(s) \approx \frac{g_m}{s C_0} \left[\frac{\dfrac{s C_0}{g_0}}{1 + \dfrac{s C_0}{g_0}} \; \frac{1 + \dfrac{s}{\omega_z}}{1 + \dfrac{s}{\omega_p}} \; \frac{1}{1 + \dfrac{s}{\omega_{eq}}} \right] \tag{4.11}$$

where ω_{eq} is the equivalent OTA parasitic pole; e.g. the pole associated with M_5 in Fig. 2.6. In this expression, g_m/sC_0 is the ideal integrator transfer function. The zero at dc and the pole at g_0/C_0 are a result of the OTA finite dc gain. For a single integrator, the dominant pole is at low frequencies, around ω_0 / A_{VOTA}. Typically, this pole-zero introduces an excess lead phase in the order of 0.2 to 2 degrees. For the OTA, the equivalent frequency of the parasitic poles and zeros is dominated by the poles, resulting in a positive frequency ω_{eq}. Hence, this pole slightly compensates the effect of the low-frequency zero-pole. If the parasitic zero is dominant, $\omega_{eq} < 0$, then excess lead phase is produced. This effect increases the phase error. For low-frequency applications, ω_{eq} can be as high as 100 times ω_0. Unfortunately, for high-frequency applications this frequency is only 10 to 20 times higher than ω_0. For example, for a 10.7 MHz filter, the parasitic poles are typically in the range of 200 MHz. It is extremely difficult to increase the frequency of the parasitic poles and zeros because large currents have to be used. Hence the OTA dc gain is reduced and some other parasitic effects become more important.

The compensating network is basically a pole-zero pair introducing some excess phase shift in the integrator [16, 21]. For high frequencies, however, the frequencies of ω_z and ω_p are not very far from the unity gain frequency of the integrator. Hence, the parasitic poles and zeros may influence the overall transfer function of the system. In this case, the OTA transconductance is a function of both the voltage-to-current converter and the compensating network. Furthermore, it should be recognized that the introduction of the pole-zero pair could increase the group delay of the integrator around the unity gain frequency. This means that the quality factor of the biquad could be increased, resulting in peaking around the resonant frequency of the filter. In high-order filters this effect increases the pass-band ripple. These effects are reduced if the frequency of the pole-zero pair is far from the resonant frequencies. The drawback is that the tenability range is further reduced as well. This is because the higher the frequencies of the pole-zero pair the lower their effect on the phase of the OTA at the resonant frequency is.

Characteristics of the PLL based techniques. Regarding the discussion of the PLL based tuning techniques, some general conclusions can be addressed.

1) These techniques are frequency analysis based techniques. In the filter based technique, a clock signal with a frequency equal to the resonant frequency of the band-pass filter is employed. The phase and the amplitude response of the filter are used for the extraction of the voltage controlled filter parameters. From the comparison of this information with the amplitude and phase of the incoming signal, the parameters of the slave filter are corrected. In the case of the VCO based approach, the

oscillating frequency and the amplitude of the oscillator output are the used parameters.

2) The best precision of the tuning system is achieved when a replica of the main filter is used in the controller. It is preferred to use a clock frequency equal to the pole frequency intended to tune. In this case, the precision of the tuning system is only limited by two factors: the mismatches between the master and slave filters and by the tolerances of the parasitic capacitors. For better matching between the master and the slave filter, it is very desirable to layout these systems as close as possible. Furthermore, the parasitic capacitors have to be accounted when dimensioning the poly-silicon capacitors.

3) Due to the unavoidable coupling capacitors, the signals due to the tuning system are coupled to the slave filter. Hence, several harmonic components due to the controller appear at the output of the filter. Obviously, the major harmonic component is at the fundamental frequency used in the controller. Because the frequency of this component is just at the poles frequency, it is in the filter pass-band. Very often the tuning system noise feedthrough is even higher than the noise level of the filter itself, mainly for high-frequency applications [13-16, 21, 33].

4) The problem of the tuning signals feedthrough can be partially overcome if the operating frequency of the tuning system is outside the filter pass-band. Nevertheless, the precision of the system is sacrificed. For instance, if the biquadratic loop is tuned at a frequency twice the pole frequency, the phase shift of the loop is tuned at that frequency too. The lead-lag network is adjusted at that frequency and the transconductance of the OTA is such that the frequency of the poles is matched with that frequency. However, there is no guarantee that the effects of the lead-lag phase network will tune the phase shift of the filter when g_m/C is scaled down. Certainly, the effects of the phase network will produce phase errors at the resonant frequencies. Hence, the precision of the automatic tuning is limited.

In the frequency based automatic tuning schemes, there is a tradeoff between precision and noise feedthrough [16]. Based on these remarks and on the analysis carried out in section 4.2, a different approach for the automatic tuning of continuous-time filters is presented in the next sections [22, 23].

4.4. Automatic Frequency Tuning technique based on charge comparison.

In this section, the principle and the implementation of a charge comparison based frequency controller is discussed. The design considerations for the control loop are discussed as well. The practical limitations of the technique are pointed out. This technique has been implemented in a 1.5 μm N-well CMOS process for the control of a 10.7 MHz band-pass filter. The results will be discussed in the section 5.4 of chapter 5.

4.4.1. The principle of the automatic frequency controller. In section 4.2, it has been demonstrated that the resonant frequency of the biquad is lowly sensitive to the OTA finite parameters. If the parasitic poles and zeros are placed at frequencies higher than 10 ω_0, they produce ω_0 errors lower than 0.5 %. Using cascode output stages, the effects of the OTA finite dc gain are also quite small. In this case the resonant frequency of the biquadratic loop is given in table 4.1. For the biquad this expression can be rewritten as

$$\omega_0 = \frac{g_{m2}}{C_1}\sqrt{\frac{g_{m3}}{g_{m2}}\frac{C_1}{C_2}} \tag{4.12}$$

This equation suggests that the frequency of the pole can be adjusted by tuning a simple g_{m2}/C_1. The ratio of the transconductances of the same order of magnitude can be accurately controlled in currently running CMOS technologies. The ratio of capacitors is quite accurate, provided that the parasitic capacitors are accounted for. Even more, the transconductance(capacitor) ratio sensitivity of ω_0 is 0.5. Hence, 1% mismatches in the transconductance(capacitor) ratio produces a 0.5 % error in ω_0.

In CMOS technologies, the parameters that can be controlled more accurately are the capacitor ratios and the dc current source ratios. The high precision of the capacitor ratios is the main reason of the success of switched-capacitor signal processors. The dc current sources have been used in digital-analog and analog-digital converters [28, 30]. Because the absence of ac-signals in the dc current source, large dimensions and large bias voltages can be used. Also, the dc current sources can be laid out inter-digitized. Replicas of a unit transistor for the current source ratios can be used instead of using transistor dimension ratios. These facts improve further the accuracy of the dc current source ratios. Using these techniques, the resolution of the current source ratios can be higher than 8 bits.

Charge injection and charge extraction. Regarding the previous observations, it is desirable to control the charge and discharge of the capacitors using a different frequency range than the filter pass-band. The charge of a capacitor can be controlled using either voltage or current. These techniques are illustrated in Fig. 4.6.

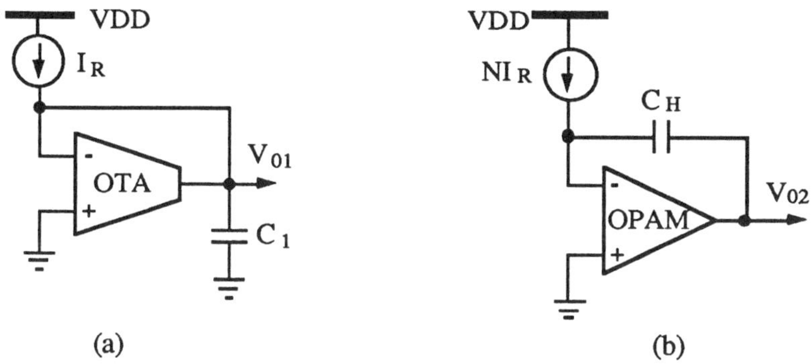

Fig. 4.6. Controlling the capacitor charge. a) charging C_1 and b) discharging C_H.

For the voltage charged capacitor, the stored charge in the capacitor can be obtained from the following expression

$$V_{01}\,C_1 = \frac{C_1\,I_R}{g_m + g_0} = \frac{C_1}{g_m}\left(\frac{I_R}{1 + \dfrac{1}{A_{VOTA}}}\right) \tag{4.13}$$

Where g_m and g_0 and A_{VOTA} are the transconductance, the output conductance and the dc gain of the OTA, respectively. The effects of the OTA finite dc gain can be considered as tolerances of the dc current source. For the reduction of this effect, the finite dc gain of the OTA has to be larger than 200. Hence, the error in the capacitor charge is of the order of 0.5%.

For the current discharged capacitor, if the initial conditions of the capacitor are zero then the time varying capacitor charge is determined by the following expression

$$V_{02}\,C_H = -\,\frac{N\,I_R}{1 + \dfrac{1}{A_{VOP}}}\,t \qquad (4.14)$$

with A_{VOP} the dc gain of the OPAMP. It is interesting to compare expressions 4.13 and 4.14. In relationship 4.13, neglecting the OTA finite dc gain effects, the charge of the capacitor is inversely proportional to the frequency of the biquad poles (see expression 4.12). Hence, if the charge of this capacitor is controlled, the poles of the second-order filter can be tuned. On the other hand, for the current discharged capacitor, the time varying charge is proportional to the injecting current. The ratio of expressions 4.14 to 4.13 is a function of both g_m/C_1 and the ratio of the current sources. Therefore, if the comparison is done in one time period, this ratio must be proportional to this period.

The frequency controller. The charge comparison of discrete packages of charge and continuous-time charge can be carried out with the circuit shown in Fig.4.7 [22]. In this charge based controller, the capacitor C_1 is charged to V_{01} during the clock phase ϕ_1. This charge is transferred to the holding capacitor C_H during the clock phase ϕ_2. Ideally, when the clock ϕ_2 is closed, C_1 is discharged producing a negative jump at the output of the OPAMP. The jump results because the discharge of C_1 is almost instantaneous. However, the jump is negative because the charge is injected into the negative plate of the capacitor C_H. At the same time, a constant current is extracted from the same capacitor plate. This current extraction results in a positive slope ramp behavior at the OPAMP output. Ideally, the total charge of the holding capacitor becomes

$$C_H V_{02}(t) = \int_{t_0}^{t} (NI_R)\,dt - (C_1\,V_{C1}(t_0))_{\phi_2} + C_H V_{02}(t_0) \qquad (4.15)$$

with V_{02} (t_0) the initial condition of the OPAMP output voltage. The integral corresponds to the extracted charge by the current source NI_R. The second term is the injected charge, during ϕ_2, from the capacitor C_1 into C_H . If the time constants are sufficiently large to charge and discharge C_1, the total injected charge is given by the relationship 4.13.

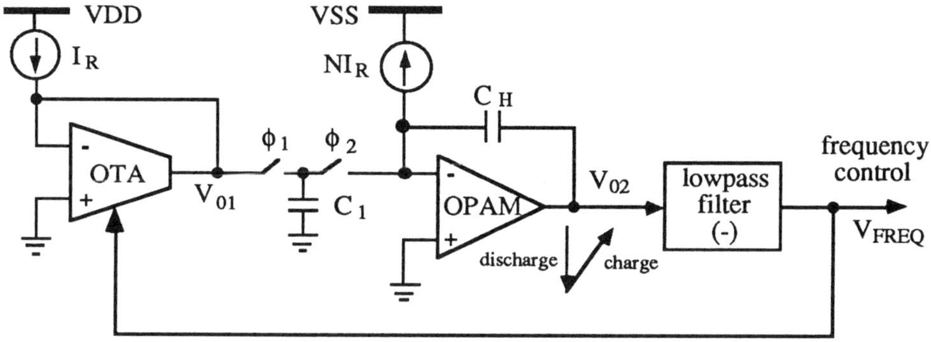

Fig. 4.7. Block diagram of the charge based automatic frequency controller.

In the loop, the OPAMP output voltage V_{02} corresponds to the error voltage. The dc value of the error voltage is inverted and filtered out by the low-pass filter. This value corresponds to the frequency control voltage, V_{FREQ}. This control voltage is fed back to control the bias current of the OTA. If the error voltage V_{02} increases the control voltage V_{FREQ} decreases. Hence, the transconductance of the OTA decreases resulting in larger V_{01}. Therefore, the charge injected by C_1 increases, resulting in an increment of the negative jump. As a consequence of this, the OPAMP output voltage V_{02} is decreased. The contrary occurs if the voltage V_{02} decreases. In steady state the average of the error voltage is constant. Hence, in a clock period the injected charge is equal to the extracted charge. In expression 4.15, however, this is equivalent to $V_{02}(T+t_0) = V_{02}(t_0)$, with T the clock period. If the effects of the OTA and OPAMP finite dc gains are neglected and equating the injected and the extracted charge in one clock period, from equations 4.12-4.15, the steady state condition of the loop can be expressed as

$$\frac{g_m}{C_1} = \frac{1}{NT} = \frac{f_{clock}}{N} \tag{4.16}$$

with N the ratio of the dc current sources. It is evident that using this technique g_m/C_1 can be tracked to the clock frequency. The factor N increases further the versatility of the system. This factor is the result of the weighted current sources used. Using N < 1/6.28, the operating frequency of the loop is lower than the poles frequency. In band-pass filters, most of the noise feedthrough due to the tuning system falls in the stop band of the filter. If N > 1/6.28, the operating frequency of the control loop is higher than the frequency of the poles. For low-pass filters, this approach could be used. However, in this case the system has to be slightly modified. Because the relatively high output impedance of the OTA, if the time duration of the clock phase ϕ_1 is shorter than 5 C_1 / g_m, the capacitor C_1 is not completely charged to V_{01}. The problem can be solved using a buffer at the V_{01} output. A couple of solutions are described in appendix A 4.1.

4.4.2. Realization of the automatic frequency controller based on charge comparison. A fully-differential block diagram of the frequency controller is depicted in Fig. 4.8. In this diagram grounded capacitors are used but obviously floating capacitors can be included as well. The OTA1 makes the conversion from the control voltage V_{FREQ} to a voltage inversely proportional to the OTA1 transconductance. This voltage is converted to charge through the capacitors C_1. The charge of these capacitors are injected into the capacitors C_H. Due to the differential nature of the switched-capacitor integrator, the behavior of the charge comparator is lowly sensitive to the common-mode offset of both OTA1 and OPAMP. The differential OPAMP output voltage is converted to a single-ended signal by the OTA and the active load resistor. This resistor is implemented by two "diode connected transistors" biased by the current source I_B. The transistor array leads to a load resistor equal to $2/g_{m7}$. The advantages of this differential-to-single-ended conversion will become apparent shortly. The low-pass filter can be implemented with a simple RC network [12-14] or using a switched-capacitor low-pass filter, as shown in the Fig. 4.8 [22]. For the RC filter, there is a tradeoff between the cutoff frequency of the low-pass filter and the settling time of the system. In the next section it is shown that this is not the case if the switched-capacitor filter is used.

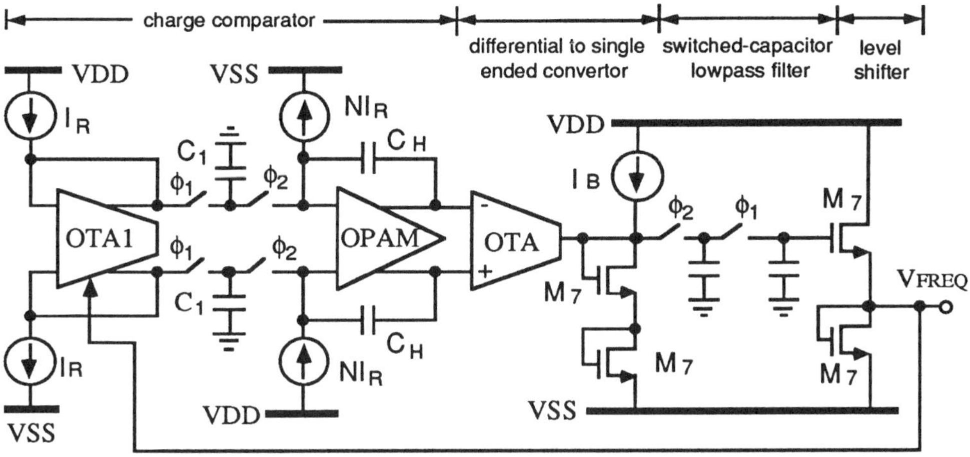

Fig. 4.8. Charge based fully-differential automatic frequency controller.

At the end of the controller, the filtered voltage is level shifted by the voltage divider. It can be noted that the last level shifter can be avoided. In fact, the level shifter is already present in the load impedance of the differential-to-single-ended converter. However, in that case the output impedance of the controller should be capacitive because the low-pass filter. Because this voltage is used for several OTAs, it is very desirable to have a low impedance node at the output of the controller. Hence, either a buffer or a current mirror has to be used, resulting in a more complex realization.

Some simulated results. SPICE simulations for the loop, at start up, are depicted in Fig. 4.9. In this figure, the OPAMP output voltages are shown. The OPAMP output voltages have a saw-tooth waveform. The jump is due to the injected (extracted in the other OPAMP input) charge from C_1 and the ramp is due to the extracted (injected) dc current NI_R. The voltage V_{FREQ}, which controls the bias current of OTA1, is shown in the bottom trace. Due to the switched-capacitor low-pass filter, this voltage has a sampled and held behavior. The clock frequency is 15 kHz and the frequency of the tuned pole is 100.2 kHz. The ideal value for factor N is 1/42. Simulated results, including 1% transistor mismatches, have shown a pole frequency error in the order of 1 %. Similar results have been obtained for the a 10.7 MHz band-pass filter.

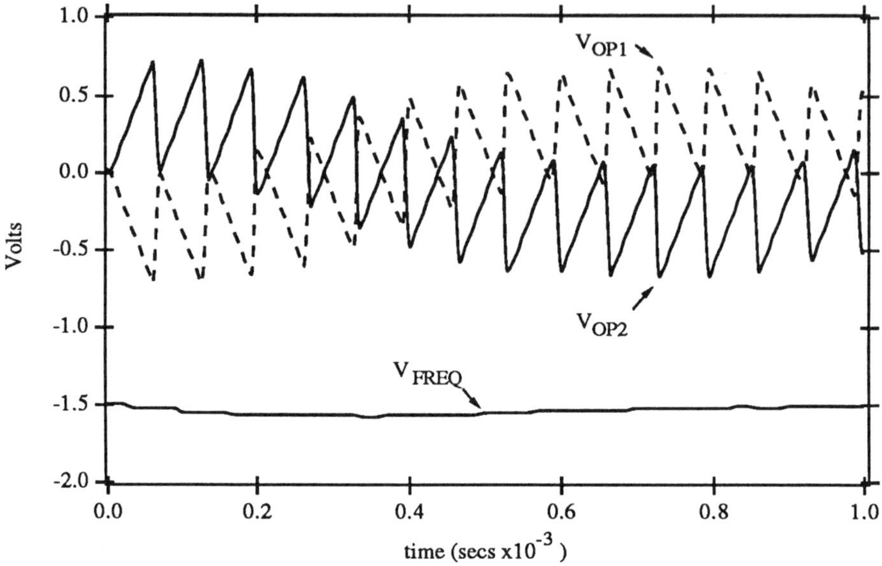

Fig. 4.9. Simulated results for the frequency tuning loop. The outputs of the OPAM and the frequency control voltage are shown in the top and bottom trace, respectively.

The low-pass filter. The low-pass filter of the automatic frequency control shown in Fig. 4.8 can be implemented in several ways. The simplest one is a RC voltage divider with the following transfer function

$$H(s) = H_0 \frac{1}{1 + \dfrac{s}{\omega_{0LP}}} \tag{4.17}$$

with H_0 and ω_{0LP} the dc gain and the cutoff frequency of the filter, respectively. In order to reduce further the harmonic components of the control signal, V_{FREQ}, the cutoff frequency of the low-pass filter has to be much lower than the clock frequency. The dominant harmonic component of the OPAMP output is at a frequency equal to the clock frequency. If a first-order RC filter is employed, the harmonic component present at the output of the low-pass filter can be approximated by

$$V_{FREQ} \approx H_0 \left(V_{02D} + \frac{f_{0LP}}{f_{clock}} e^{j\varphi} v_{02d} \right) \qquad (4.18)$$

where V_{02D} and v_{02d} are the dc voltage and amplitude of the frequency component at f_{clock} of the OPAMP output voltage, respectively. In the relationship 4.18, f_{clock} (= 1/T) is the clock frequency and φ is the v_{02d} phase shift due to the low-pass filter.

According to 4.18, the lower the cutoff frequency of the low-pass filter, the higher the rejection to the tuning system noise is. For the reduction of this noise it is desirable that $f_{0LP}/f_{clock} < 0.01$. The design of this filter is quite easy if an external capacitor is used. However, for on-chip realizations a large passive resistor has to be employed.

In Fig. 4.8, the low-pass filter has been implemented using a switched-capacitor based resistor. This approach has several desirable properties that improve the efficiency of the system, mainly for full on-chip applications. Firstly, there is no penalty for the use of the switched-capacitor resistor because the clock phases are already available. The silicon area needed for the switched-capacitor resistor is quite small. In switched-capacitor circuits large resistors requires minimum silicon area, which is not the case for passive resistors. Secondly, another very important advantage of the switched-capacitor realization is its sampled nature. In the next section is shown that if the clock phases are arranged as shown in Fig. 4.6, the sampled voltage is just the dc voltage of the OPAMP output voltage. In addition, the clock frequency of the switched-capacitor low-pass filter is at the same frequency that the major harmonic component generated by the charge comparator. Hence, the clock frequency and its harmonics are eliminated by the switched-capacitor filter. This is the reason of the absence of harmonic components in V_{FREQ} (see Fig. 4.9). However, the switched-capacitor filter introduces clock-feedthrough offset voltage. This offset voltage can easily be reduced by well known techniques [25].

Properties of the loop. There are several interesting properties of the loop that makes this option attractive for the automatic tuning of continuous-time signal processors. Some of them are discussed in this section.

1) The circuit is based on charge comparison. The first component of the charge is due to the current controlled voltage source, the unity feedback connected OTA1. This charge is periodically injected into the hold capacitor, C_H. This charge is proportional to the time constant, C_1/g_m, intended to control. In addition, the proportionality constant is a dc current source. The second charge component is the

result of a continuously extracted dc current. Because both charges are proportional to a dc current sources, the tuned time constant is proportional to the ratio of the dc current sources. The other parameter involved in the charge comparison is the clock frequency. As a result, both parameters, the dc current sources ratio and the clock frequency, can be used for the tuning of the filter time constant. This fact allows to operate the loop at a different frequency range than the filter pass-band frequency range. Hence, most of the noise tuning feedthrough falls in the stop band of the filter. The result is that the filter dynamic range is almost not affected by the operation of the automatic tuning system.

2) The accuracy of the system is further analyzed in the next section. Nevertheless, it can be noted that this accuracy depends of the comparison between the extracted and the injected charge into the capacitor C_H. In steady state, the injected and extracted charges are equal. In the ideal case, this comparison is independent of C_H. The capacitance C_H is used as an auxiliary element in which the charge is injected and extracted. In a first approximation, this capacitor is just a scaling factor in the behavior of the system.

3) For the control of the loop, the dc signal of the OPAMP output voltage is used. This signal is controlled by the charge comparator. If the system is biased in its linear operation region, the precision of the tuning system is independent of this signal. As it was mentioned in 2), the accuracy of the system depends of the charge comparison instead of the dc output voltage of the OPAMP. This is a very important property of the loop. Hence, the accuracy of the system is lowly sensitive to the offset voltage of both the differential-to-single-ended converter and the low pass filter. For the same reason, the precision of the level shifter is not critical in the behavior of the system.

4) After the OPAMP, the differential signal is converted to a single-ended signal. It has to be noted that an OTA with a large linear range has to be used for the differential-to-single-ended conversion. If there is a large dc output voltages at the output of the OPAMP then the OTA has to be able to detect this voltage. If the switched-capacitor is used, however, then it is not necessary to detect accurately the whole OPAMP output voltage range. In this case, it is sufficient to convert accurately the OPAMP output voltage-to-current at the end of the clock phase at which the switched-capacitor circuit is sampling the signal.

5) For the control of the loop, only the dc signal is used. This signal can be level shifted in order to optimize the dynamic range of the loop. For this purpose, the

active load impedance of the OTA is implemented with the transistors and the current source I_B (see Fig. 4.8). It can be noted that those transistors are equal to the transistors used for the biasing of the OTA1 (e.g. see Fig. 2.4 in chapter 2). However, the bias current of the active impedance is equal to the OTA designed bias current. In addition, the transistors used in the level shifter are similar to these used in both OTA active impedance and OTA1. Under these conditions and if the variations of the process parameters are small, the dc current sourced for the OTA should be very small in steady state. Hence, for stable operation the dc signal at the output of the OPAMP is very small. The result of this is an almost optimum operation point for the frequency controller.

There are several second-order effects present in the system. For instance, the finite dc gain of both OTA1 and OPAMP. These effects among others are considered in the next section.

4.4.3. Design considerations for the charge based frequency controller. The mismatches in dc current ratios can be as small as 0.4 % (8 bits) and sometimes better [27-30]. So, if properly designed, the precision of the factor N (dc current sources ratio) can be quite good. The reduction of the mismatches in the current sources is a matter of the layout and symmetries. The effects of the gate area, bias voltage and orientations of the transistors, has been already mentioned. Further details of these effects can be found elsewhere [27 - 29].

The major design problem of this loop is the mismatch between the filter capacitors and the capacitors associated with the automatic tuning loop. In the filter, there are several parasitic capacitors due to the additional devices connected to the output of the OTAs. A part of these parasitic capacitors are the drain capacitors of the OTA output stage and their value is not very well controlled. Another parasitic capacitor component is due to the gate-source and gate-drain capacitors of the OTA input stage. The top capacitors of the poly-silicon capacitors have to be considered as well. Carefully modeling is necessary, from the layout dimensions, and simulation of these capacitors in order to include their effect in C_1. This is basically a pre-distorting procedure for the poly-silicon capacitors. Unfortunately, this is a manual procedure that requires extensive simulations. Even more, the mismatches can be minimized but they can not be eliminated because of process parameter variations. Certainly, the matching between the controller and the slave filter improves if the poly-silicon capacitors are much larger than the parasitic capacitors. However, increasing the OTA bias current the OTA finite dc gain is reduced. In the next section it is demonstrated that the OTA finite dc gain reduces the accuracy of the system. Therefore, there is a tradeoff between the value of the poly-silicon capacitors and the power consumption and silicon area. Hence, the higher the value of these capacitors the better the matching, but the OTA transconductance has to be increased as well. The result is an increment in both power consumption and silicon area.

Effects of the OTA1 and OPAMP second-order parameters. In order to make clear the analysis, a single-ended version of the controller is discussed. Nevertheless, the same analysis applies to the fully-differential version. A simplified diagram of the charge comparator is shown in Fig. 4.10 (see Fig. 4.7).

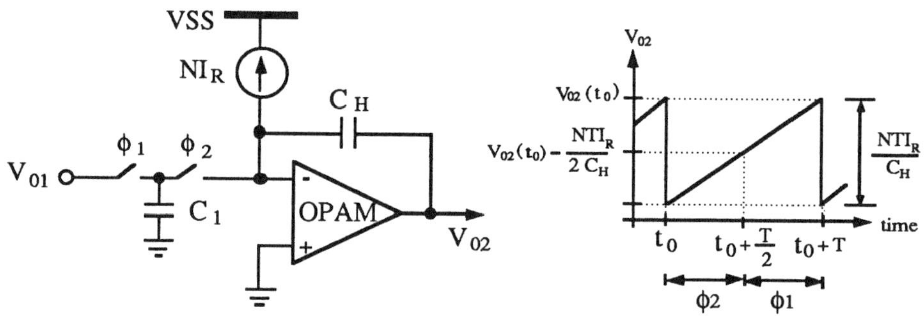

(a) (b)

Fig. 4.10. a) Simplified diagram of the charge comparator.
b) The OPAMP output voltage and the clock phases.

The clock phases involved in the system are defined as follows:

$$\phi_2 \rightarrow t_0 < t \leq t_0 + \frac{T}{2} \qquad \phi_1 \rightarrow t_0 + \frac{T}{2} < t \leq t_0 + T$$

with T the clock period. During the clock phase ϕ_2, the variation of the C_H charge is given by the difference between the injected charge and extracted charge due to the current source NI_R and the capacitor C_1, respectively. The charge variation on the capacitors can be expressed by the following charge conservation equation:

$$C_H\left(v_{CH}(t) - v_{CH}(t_0)\right) = C_1\left(v_{C1}(t) - v_{C1}(t_0)\right) + N\,I_R\left(t - t_0\right) \tag{4.19}$$

where v_{CH} is the voltage at the plates of the capacitor C_H. If the OPAMP dc gain is equal to A_{VOP}, the OPAMP output voltage can be expressed as $v_{02}(t) = A_{VOP}(V_+ - V_-)$, with V_+ and V_- the non-inverting and inverting OPAMP input voltages, respectively. In 4.19, the voltage v_{CH} can be expressed as a function of the OPAMP output voltage, v_{02}, and A_{VOP} as

$$v_{02}(t) = \frac{\left(1+\frac{1}{A_{VOP}}\right) C_H v_{02}(t_0) + C_1 v_{C1}(t_0) + N I_R (t - t_0)}{C_H + \frac{1}{A_{VOP}}(C_H + C_1)} \tag{4.20}$$

During the next clock phase, ϕ_1, the capacitor C_1 is connected to OTA1 (see Figs. 4.7 and 4.8). If the OTA1 output conductance is accounted, the charge of the capacitor C_1 can be expressed by expression 4.13. In addition, the dc current source NI_R extracts a constant current NI_R from the capacitor C_H. Due to the extracted current, the OPAMP output voltage presents a saw-tooth waveform (see Fig. 4.10b). Equating the dc current with the capacitor current and taking into account the OPAMP finite dc gain, the computation of the OPAMP output voltage during ϕ_1 yields

$$v_{02}(t) = v_{02}\left(t_0 + \frac{T}{2}\right) + \frac{N I_R \left(t - t_0 - \frac{T}{2}\right)}{1 + \frac{1}{A_{VOP}}} \tag{4.21}$$

If equation 4.21 is evaluated at the end of ϕ_1 ($t = t_0 + T$) and the result is inserted in expression 4.20 then the OPAMP output voltage at the end of ϕ_2 becomes

$$v_{02}\left(t_0 + \frac{T}{2}\right) = \frac{\left(1+\frac{1}{A_{VOP}}\right) C_H v_{02}\left(t_0 - \frac{T}{2}\right)}{C_H\left(1+\frac{1}{A_{VOP}}\right) + \frac{C_1}{A_{VOP}}} + \frac{C_1 v_{C1}(t_0)}{C_H\left(1+\frac{1}{A_{VOP}}\right) + \frac{C_1}{A_{VOP}}} + \frac{N T I_R}{2 C_H} \tag{4.22}$$

The steady state condition of the loop is reached when $v_{02}(t_0+T/2) = v_{02}(t_0-T/2)$. If equation 4.13 is inserted in 4.22 and if the third-order terms are neglected, the steady state condition for the control loop can be expressed by the following relationship

$$\frac{g_m}{C_1} = \frac{1}{NT}\left(1 - \frac{1}{A_{VOTA}} + \frac{1}{A_{VOP}}\,\frac{C_1 V_{02}\left(t_0 + \frac{T}{2}\right)}{N T I_R}\right) \tag{4.23}$$

Therefore, the tuning error is approximately given by

$$\text{Error} \approx -\frac{1}{A_{VOTA}} + \frac{1}{A_{VOP}}\,\frac{C_1 v_{02}\left(t_0 + \frac{T}{2}\right)}{N T I_R} \tag{4.24}$$

In this equation, the effects of the finite dc gain of both the OTA and the OPAMP can be distinguished. The OTA produces an error inversely proportional to its finite dc gain. The effects of the OPAMP finite dc gain are accounted in the second term of equation 4.24. In this term, the voltage $v_{02}(t_0+T/2)$ is the dc value of the OPAMP output voltage (see Fig. 4.10b). The physical reason of this term is explained as follows: The charge of C_1 is injected into C_H at the beginning of the clock phase ϕ_2. This charge produces a negative jump in the OPAMP output voltage. Because the continuously extracted charge NI_R, the voltage increases with the time. Due to the finite dc gain of the OPAMP, there is a charge error at the input of the OPAMP equal to $C_1 V_{02}(t) / A_{VOP}$. At the end of the clock phase ϕ_2, the capacitor C_1 is disconnected from the OPAMP. The charge of C_1 at the end of ϕ_2 is not injected to C_H and so it is the responsible for this tuning error. In the worst case, the error due to the OPAMP finite gain is equal to $1/A_{VOP}$. Thus, using a high dc gain OPAMP this error is further reduced.

If the dc voltage of the OPAMP output voltage $v_{02}(t_0+T/2)$ is zero, then the tuning error due to finite dc gain of the OPAMP vanishes. This is because in the instant in which C_1 is disconnected from the OPAMP its charge is zero. Hence, both the charge of C_1 and the charge of the current source are completely sourced and sinked to the capacitor C_H, respectively. In this case, the only tuning error is due to the OTA finite dc gain.

Effects of the OTA and OPAMP offset voltages. The dc offset voltage of the OTA, see Fig. 4.7, produces tuning errors as well. The OTA offset voltage effects impose the lowest limit to the voltage V_{01}. In addition, the offset voltage of the OPAMP introduces a tuning error. If both OTA and OPAMP offset voltages are accounted for, and if the effects of the dc gain of the OTA and the OPAMP are neglected, it can be shown that the tuning error is approximately given by

$$\text{Error} \approx \frac{V_{\text{off-OTA}} - V_{\text{off-OPAMP}}}{V_{01}} \tag{4.25}$$

with $V_{off\text{-}OTA}$ and $V_{off\text{-}OPAMP}$ are the differential offset voltages of the OTA and OPAMP, respectively. For instance, if the total dc offset voltage is around 5 mV, for an accuracy of 1% V_{01} has to be larger than 0.5 Volt. For 0.5 % accuracy, this voltage should be larger than 1 V. For fully-differential structures this is not a major limitation because the symmetry of the structure guarantees very small differential dc offsets. In the fully symmetrical implementation, the accuracy of the tuning system is little sensitive to the common-mode offset voltages, provided that the capacitor mismatches are small (e.g. less than 1 %).

Taking into account these considerations, the accuracy of the loop is in the order of 1 %, including 1 % transistor mismatches. However, the controller can follow clock variations of about ± 50 % with an accuracy still around 1%. Some results will be discussed in the next chapter.

4.5. Automatic BW controller based on transient analysis.

The BW detection of a biquadratic filter is not always an easy task. In section 4.2, it has been shown that the bandwidth of a biquad is extremely sensitive to the OTA finite parameters. These effects are particularly more severe for narrow band filters. The problem of the BW detection is complex because the transfer function is sensitive to BW only around the resonant frequency of the filter. For low and high frequencies, the behavior of the filter is dominated by the frequency of the poles. Hence, in the frequency domain the BW or Q detection requires a test signal at the resonant frequency. The result, as discussed earlier, is that the interference of the tuning signal feedthrough with the main signal increases the pass-band noise.

The time domain analysis can also be used for the BW or Q tuning of the filter. However, the tuning system is more complex. The main ideas behind this approach are:

1) The operating frequency of the BW tuning loop can be designed much lower than the resonant frequencies of the main filter. This fact reduces the effects of the noise feedthrough on the filter pass-band. Due to the lower frequency used, the processing of the detected signal is easier and the signal feedthrough is reduced, even

for low-pass filters. This is because the signal feedthrough is coupled to the filter output mainly through the coupling capacitors. Hence, the coupling is reduced at lower frequencies.

2) The bandwidth of a biquadratic filter can be considered as composed by 2 parts, as expressed in equation 4.26 (see section 4.2 for details).

$$BW = \frac{\omega_0}{Q} \cong \left(\frac{g_{01} + g_{03}}{C_1} + \frac{g_{02}}{C_2} + 2\frac{\omega_0}{\omega_{eq}}^2 \right) + \left(\frac{g_1}{C_1} + \frac{g_2}{C_2} \right)$$
$$= (BW)_{par} + \left(\frac{g_1}{C_1} + \frac{g_2}{C_2} \right) \tag{4.26}$$

The first part of the BW is due to all the parasitic effects like OTA finite dc gain and parasitic poles and zeros. The second part is the tunable part of BW. The goal of the BW controller is to adjust this part of the BW in order to compensate the parasitic effects.

3) In this approach, it is not tried to improve the performance of the OTA, which should decrease the parasitic term of 4.26. To increase the dc gain of a high-frequency OTA is extremely difficult. This probably should require the use of triple cascode stages. The use of these stages reduces the dynamic range of the OTA, introduce extra poles and the design becomes more complex. On the other hand, to push the parasitic poles to higher frequencies requires a lot of silicon area and power consumption.

4) The tuning system assumes that the parasitic term is already present. The only task of the automatic tuning system is to adjust g_1/C_1 or $g_1/C_1 + g_2/C_2$ such that the total bandwidth of the filter corresponds to the required value.

With these considerations in mind, the automatic Q tuning system based on transient analysis is described in the next sections.

4.5.1. The principle of the technique. In a biquadratic filter, the bandwidth of the filter is inversely proportional to the filter quality factor, $BW = \omega_0 / Q$. If ω_0 is adjusted using the frequency control loop, then this frequency can be considered as a fixed parameter. Therefore, the tuning of the BW implies an adjustment in the filter quality factor. Under these conditions, the Q detection and the

BW detection are correlated to each other. The Q of a second-order filter can be detected in both frequency and time domains. The Q detection in the frequency domain has been discussed elsewhere [14-15,24]. It is important to point out that this detection is carried out at the center frequency of the filter. In the time domain, the pulse response of a biquadratic filter allows also the Q detection. A discussion of this technique follows.

Transient response of the low-pass filter. In the frequency domain, the response of a first-order filter is totally characterized by two parameters. Namely, the dc gain of the single pole filter and its -3 dB frequency (pole frequency) [26]. In the time domain, the behavior of the filter can be described recognizing the impulse response of the filter. The step response of a first-order low-pass filter with unity dc gain and -3 dB frequency at ω_{-3dB} is given by

$$v_{OLP}(t - t_0) = V_{OLP}(t_0) + \left(1 - \exp^{-\omega_{-3dB} t}\right) V_{step} \tag{4.27}$$

with $V_{OLP}(t_0)$ the initial condition of the low-pass filter output. t_0 is the time at which the step is applied, V_{step} is the amplitude of the applied step and ω_{-3} dB is the -3 dB frequency of the first-order low-pass filter.

The step response of a biquadratic low-pass filter with unity dc gain and bandwidth equal to BW and Q larger than 0.5 is characterized by the following expression :

$$v_{OLP2}(t - t_0) = V_{OLP2}(t_0) + (1 - h(t - t_0)) V_{step} \tag{4.28}$$

In this relationship, the function h(t) is called the impulse response of the biquad [26]. For the biquad, this function is given by

$$h(t) = \frac{1}{\sqrt{1 - \frac{1}{4Q^2}}} \exp^{-\frac{BW\,t}{2}} \sin\left(\sqrt{1 - \frac{1}{4Q^2}}\, \omega_o t + \phi\right) \tag{4.29}$$

where ϕ is a function of Q. For large Q, the first factor of 4.29 can be approximated as unity. Hence h(t) equals an unity amplitude sinusoidal function whose amplitude decays exponentially. Comparing expressions 4.27 and 4.28 it can be noted that the step response of the first and second-order low-pass filters are similar. The basic difference is the sinusoidal function present in h(t). However, the envelope (the exponential function) of both responses are similar. Even more, if ω_{-3dB} is equal to BW/2 and Q>>1 then these envelopes are equal. Under this condition, the error voltage of the BW tuning loop can be generated by comparing the step response of both filters. The first-order filter can be considered as the "reference filter". The master filter, to be tuned, is the second-order filter.

Because the absence of loops in a first-order filter, the pole frequency of the reference filter is little sensitive to the finite parameters of the OTAs. The frequency response of the first-order low-pass filter is rather insensitive to the parasitic poles. Moreover, both the dc gain and the -3 dB frequency of a first-order filter can be accurately controlled in CMOS technologies using the frequency controller. The implementation of the complete system is discussed in the next sections.

Block diagram of the BW controller. A block diagram implementation of this technique is depicted in Fig. 4.11. In this diagram, the input signals are 2 complementary pulse trains. The pulsed values for v_p are 0 and V_p, and for $\overline{v}_p$ are 0 and $-V_p$. The implementation of the second and first-order low-pass filters will be discussed shortly. The envelope of both filters are detected by the envelope detectors. These envelopes are compared and filtered by the low-pass filter. From this comparison, the correcting voltage is generated and fed back to adjust the resistor or resistors that determine the quality factor of the biquadratic filter.

It has to be noted that the envelopes have to be compared when V_p is in the low state. This is because in the envelope detector, only the most positive signal is detected. In the positive transition the detected signal at the output of the first-order filter corresponds to the typical exponential charging response. However, the detected signal from the second-order filter corresponds to an exponential discharge starting from $2V_p$. This is a result of the sinusoidal function implicit in equations 4.28 and 4.29. Hence, the envelopes have to be compared in a single phase. This is not a major problem because the low-pass filter can be implemented as a switched-capacitor filter. So, the sample and hold operation is implicit in the filter. The active resistor controls the bandwidth of the filter (see section 4.2 and equation 4.19). It can be implemented using either of the earlier discussed topologies.

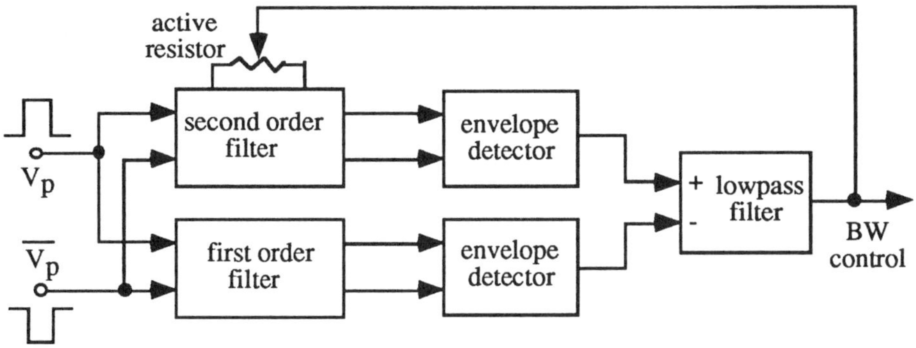

Fig. 4.11. Block diagram of the automatic bandwidth controller.

4.5.2. Realization of the automatic BW controller. The implementation of the automatic BW control involves the design of several building blocks (see Fig. 4.12). The master filter is just a replica of a main filter second-order section. It requires 3 OTAs with the same transconductance. In this system a low-pass filter is used as the master filter. Because the same OTAs used, the dc gain of the filter is unity. The quality factor of the biquad is controlled through the active resistor, R_1. In this diagram, the same capacitors, C_1, have been used for both integrators. In general, these capacitors can be different. The reference filter has been implemented with two OTARs, with the same transconductance, and the capacitor $2C_1$. The unity feedback connected OTAR guarantees the unity dc gain of the reference filter. The -3 dB frequency of the reference filter is adjusted with the load capacitor $2\ C_1$ and g_{mR}. The transconductance g_{mR} is Q times lower than g_{m1}.

The envelope detector involves a magnitude detector based on a rectifier detector [21]. It will be discussed in the next section. The output of the envelopes are compared by the OTA and the error voltage is converted into current. This current is integrated by the capacitor C_I only during the clock phase ϕ_4. This clock phase is synchronized with the negative transition of the input signals. During the high pulsed value, clock phase ϕ_3, the output of the OTA is connected to the ground. Hence, the parasitic capacitors are discharged and so have a minimum effect on the performance of the loop. The low-pass filter is implemented by the action of the OTA and the OPAMP together with the integrating capacitor C_I. The output of the OPAMP is sampled during the clock phase in which the signal is not being injected into the integrating-holding capacitor. In this way, the value of the resistor is adjusted during the "dead time" of the integrator. Therefore, due to the sample and hold action, the adjusted value remains constant during the comparison phase, clock phase ϕ_3. As in the frequency controller, the level shifter allows to optimize the operation point of the loop. In Fig. 4.12, two N transistors connected in series are used. This type of level shifter is well situated for the control of the current source based resistors, already discussed in chapter 3. Obviously, PMOS transistors can be used as well.

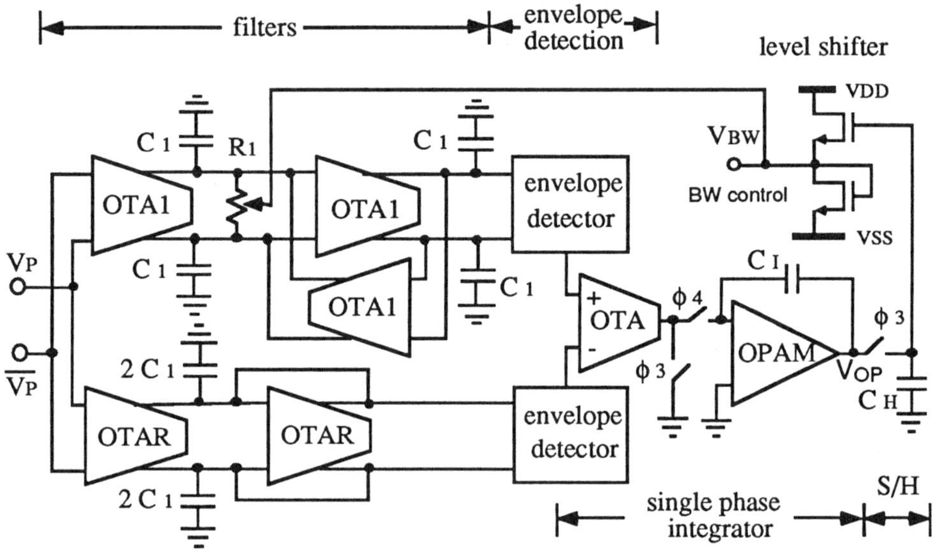

Fig. 4.12. A fully-differential version of the BW frequency controller.

The envelope detector. The envelope detector consists of three blocks (see Fig. 4.13). The first part is the amplifier and level shifter. This is implemented by the low-distortion voltage-to-current transducer and the diode connected transistors M_d. The voltage-to-current transducer with large input range is used because the large-signal present at the output of the filters. For smaller signals, a simpler topology can be used.

The signal is rectified by the differential-pair, transistors M_{det}. For better rectification, the transconductance of the M_{det} transistors have to be very high. In that case, the signal at the common-source follows the more positive input signal. The third block is composed by the current source I_{det} and the detector capacitor C_{det}. The current source discharges the output node of the detector. The current source and the capacitor determine the discharging speed of the detector.

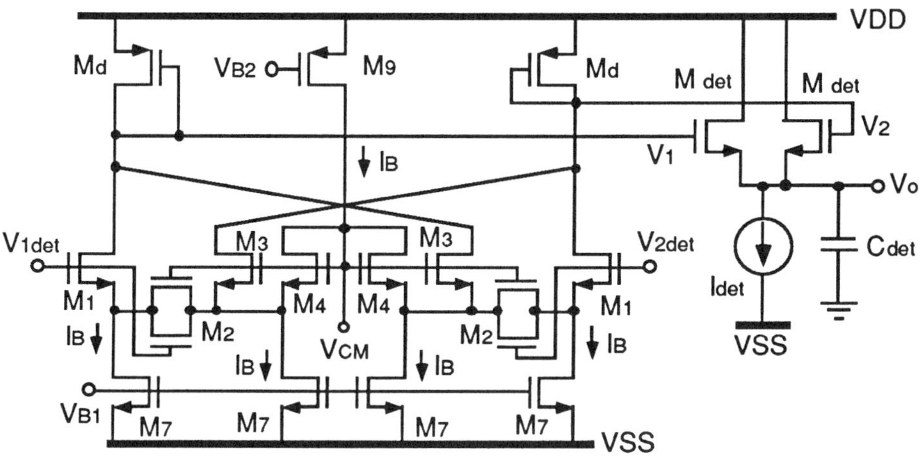

Fig. 4.13. Envelope detector using the low-distortion voltage-to-current transducer.

The charge and discharge of the envelope detector can be mathematically modeled by the following equations:

$$v_{out} = \max\left(v_1 - V_{GS}, v_2 - V_{GS}\right) \qquad \text{charging}$$
$$v_{out} = v_{out\ char} - \left(\frac{I_{det}}{C_{det}}\right)t \qquad \text{discharging} \qquad (4.30)$$

where max means the maximum between v_1-V_{GS} and v_2-V_{GS}. During the charging time, the output of the envelope detector is equal to the higher differential-pair input voltage, v_1 or v_2 , level shifted by the gate-source voltage of M_{det}. If the common-mode voltage of the differential-pair input signal decreases rapidly, the capacitor C_{det} is discharged by the dc current source. The discharge of the capacitor is time proportional with a slope given by the constant I_{det} / C_{det}.

Simulated and breadboard results. Simulated results for the loop at starting up are shown in Fig. 4.14. For this plot, the quality factor of the filter is 4. In the middle trace, referred to the left-hand side scale, the outputs of the master and the reference filters are shown. It can be noted in these traces that the pulse response of the master filter, solid traces, follows the transient response of the reference filter, soft traces.

The output of the envelope detectors are shown in the bottom traces of this figure. These traces are referred to the right-hand side scale. It can be recognized that the envelopes are adjusted only during the high to low transitions. The OPAMP output voltage is depicted in the top trace (see Figs. 4.12 and 4.14). In this trace, the effect of the injected current can be noted. If this voltage is directly fed back to the resistor, the filter becomes a time variant system. Using the sample and hold, see Fig. 4.12, the voltage applied to the resistor remains constant during the comparison clock phase. Hence, during this clock phase the filter becomes approximately a time invariant system.

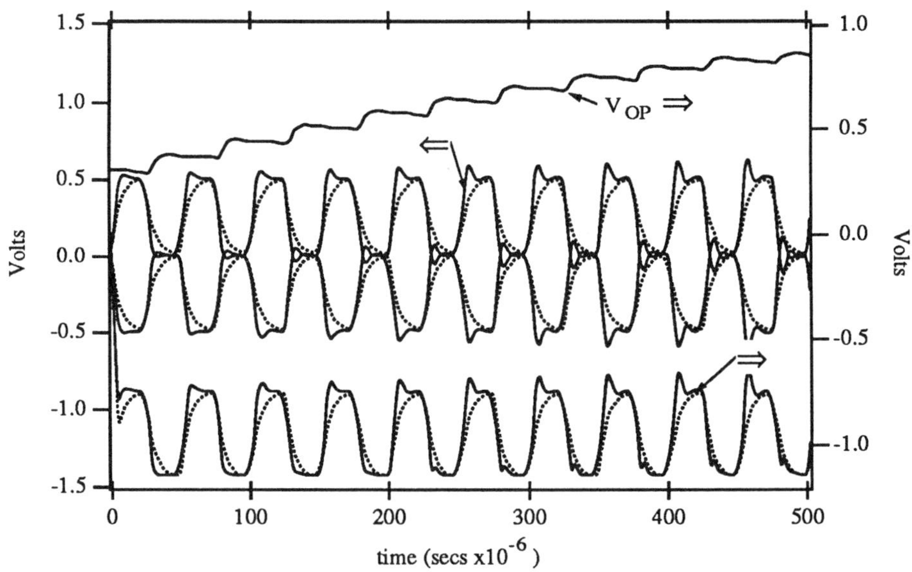

Fig. 4.14. Simulated results for the Q tuning loop. The outputs of the master and the reference filters are shown in the middle trace. The output of the envelope detectors and the bandwidth control voltage V_{OP} are shown in the bottom and top trace, respectively.

When the control voltage reaches its stable state, the transient response of both filters are adjusted. It can be noted in the middle traces that the envelope of both filters are adjusted after the tenth clock cycle. Typically, the system converges to its final value after 8 to 10 clock cycles. From simulations, the error in the absolute value of the tuned Q, including 1% transistor mismatches, is about 5 to 10 %.

A single-ended breadboard using discrete OPAMPs, discrete OTAs, 1% tolerance resistors and 5% tolerance capacitors has been implemented. A CMOS transistor biased in triode region has been used as a voltage controlled resistor. The experimental results for two cases are shown in Fig. 4.15. The top traces are related to the right-hand side axis. In this plot, the output of both envelope detectors are depicted. For this case, the ideal quality factor of the biquad is Q=15. The bottom traces are similar to the top traces but for Q=2.5. The bottom traces are related to the left-hand side axis.

In Fig. 4.15 when the applied pulse is high the envelope detector associated with the first-order filter, soft traces, follows the classical exponential response. However, the envelope detector associated with the biquad follows the positive envelope of the filter response. In this clock phase, the transient response of the filters are not compared. The envelopes are only compared when the applied signal is low. In this case, both envelopes follow the exponential response. It can be seen in this plot that both transient responses are quite similar during the high to low transition. Therefore, the master filter is tracked to the reference filter.

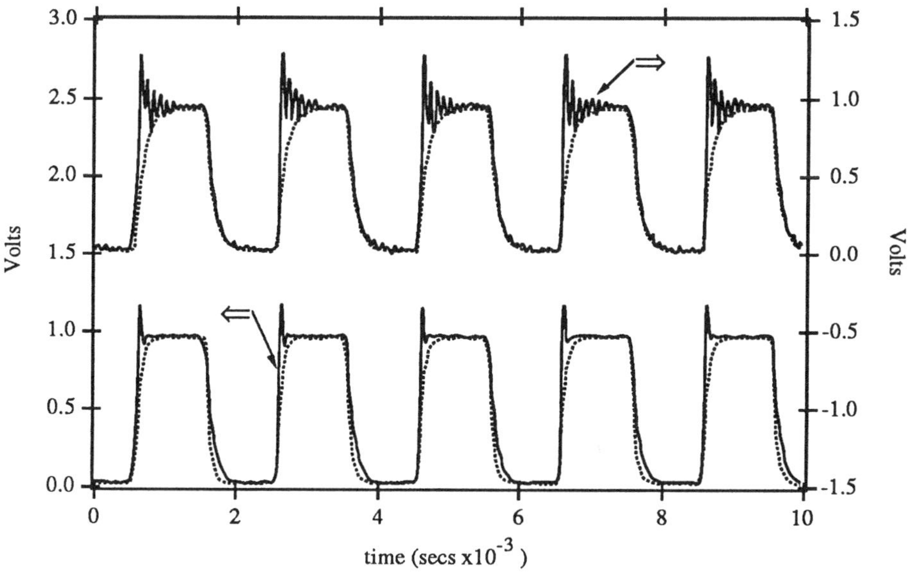

Fig. 4.15. Breadboard results. Envelope detector outputs for two cases. The top trace corresponds to Q=15 and the bottom trace to Q=2.5.

4.5.3. Design considerations for the BW controller. In this section, several design considerations for the BW tuning loop are discussed. The effects of the OTAs and OPAMPs finite parameters on the accuracy of the loop are considered. It is shown that the accuracy of this loop is limited. The loop is sensitive to the offset voltages of the active blocks. However, it is shown that the precision of the loop is sufficient for most of the practical applications. A discussion of the major design considerations follows.

The BW tuning loop. In the BW tuning loop, the pole of the first-order reference filter has to be equal to half the bandwidth of the biquad, that is equal to $\omega_0/2Q$. If the value of the capacitor associated with the reference filter is twice the value of the capacitor in the frequency control loop, then the transconductance associated with the reference filter, g_{m1}, has to be equal to g_{m2}/Q. Then, for high Q filters large transconductance ratios are required. Unfortunately, the large transconductance ratios are not quite well controlled in current CMOS processes. These large ratios very often depend of both effective transistor width ratios and effective transistor length ratios. The uncertainty of both effective lengths and effective widths (due to lateral diffusion, birds beak, depletion regions, etc.) are two of the most limiting factors in the accuracy of these ratios. Simulated results have shown transconductance ratio errors in the order of 1 to 5%, depending of the minimum dimensions used and depending of the g_m ratio. This is a limitation of this approach. Using OTA transistors with minimum length of 6 μm and a minimum width of 4.5 μm, the precision of large OTA transconductance ratios, lower than 30, is of the order of 1 to 3 %.

Effects of the OTA offset voltage. The BW tuning loop is based on the comparison of the envelope of both reference and master filter. From the comparison of the envelopes, the error signal is generated. This signal is converted to current by the OTA (see Fig. 4.9). Because this is a low-frequency signal, the offset voltage of the OTA produces some tuning error. The situation is illustrated in Fig. 4.16.

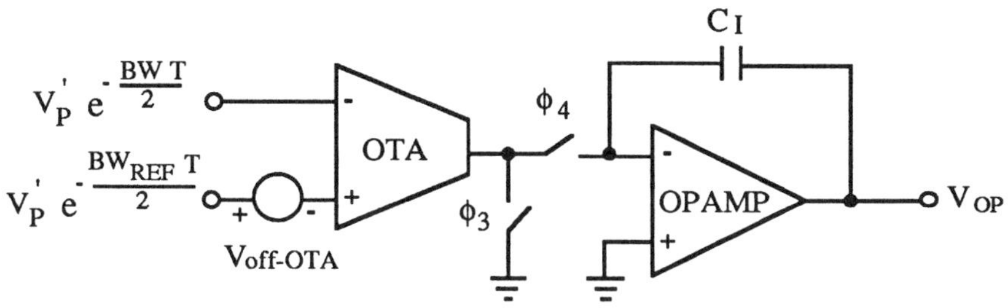

Fig. 4.16. The single phase integrator of the BW controller.
The OTA dc offset voltage is shown.

From Fig. 4.16, the OTA input voltage is given by

$$v_+ - v_- = V_P' e^{-\frac{BWt}{2}} - V_P' e^{-\frac{BW_{REF}\,t}{2}} + V_{off\text{-}OTA} \tag{4.31}$$

where V_p' is the amplitude of the envelope detector output. BW and BW_{REF} are the biquadratic filter bandwidth and the bandwidth of the reference filter, respectively. In the OTA, v_+-v_- is converted to current. During ϕ_4, the OTA output current is integrated by the OPAMP and the integrating capacitor C_I.

Since the charge of the integrating capacitor is not affected during the clock phase ϕ_3 then the OPAMP output voltage remains constant. The steady state condition of the loop is reached when the average of the OTA output current injected to the integrating capacitor is zero. From 4.31, this condition is expressed by the following equation

$$\int_{\phi_4} \left(V_P' e^{-\frac{BWt}{2}} - V_P' e^{-\frac{BW_{REF}\,t}{2}} + V_{off\text{-}OTA} \right) dt = 0 \tag{4.32}$$

The integration of this equation leads to

$$\frac{1 - e^{-\frac{BWT}{4}}}{BWT} - \frac{1 - e^{-\frac{BW_{REF}\,T}{4}}}{BW_{REF}\,T} = \frac{V_{off\text{-}OTA}}{4\,V_P'} \tag{4.33}$$

with T the clock period. For the general solution of this equation, numerical methods have to be used. If small BW deviations are assumed, the solution of 4.33 can be approximated by the following analytical solution

$$BW \cong BW_{REF}\left(1 - \frac{V_{off\text{-}OTA}}{4\,V_P^{'}}\,BW_{REF}\,T\right) \tag{4.34}$$

From this relationship, the BW error (in %) becomes

$$BW\ error\ (\%) \cong \frac{V_{off\text{-}OTA}}{4\,V_P^{'}}\,BW_{REF}\,T \tag{4.35}$$

In Fig. 4.17, results for the exact solution (numerical) of 4.33 and for 4.35 are shown. In this plot, the BW tuning error against the $V_{off\text{-}OTA}/V_R'$ is depicted. In the top axis, the OTA offset voltage for the case V_R'=0.3 V is shown.

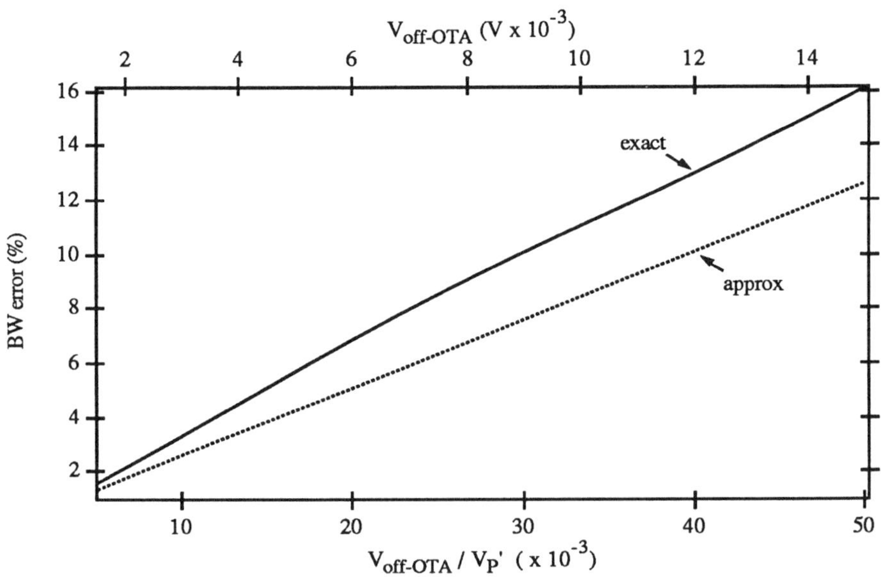

Fig. 4.17. BW tuning errors due to the OTA offset voltage.
For this plot, $BW_{REF}\,T$ = 10 and V_p' = 0.3 Volts.

According to the previous analysis, the error voltage is a function of the product $BW_{REF}\,T$. This is a normal result because the larger the comparison time the larger the effects of the integrated offsets are. However, the larger the integrating time the lower the amplitude of the envelopes are. For the detection of the envelopes it is desirable to use a clock period at least 8 times higher than BW_{REF}. The other important factor determining the accuracy of the loop is the ratio of the OTA offset voltage to V_p'. In the case that this ratio is 1/100 and considering $BW_{REF}\,T = 8$, the tuning error is of the order of 2.0 %. For this case, if V_p' is 0.3 Volts, then the OTA offset voltage has to be lower than 3 mV. A more realistic value for the OTA offset voltage is around 10 mV. In this case the tuning error is around 7 %. For $BW_{REF}\,T =$ 10, the theoretical tuning error is around 10 %. It is clear that the BW tuning errors are much higher than the typical 1 % tuning error of the frequency controller.

At this time, there are some remarks and also some important questions. The BW tuning system is more complex than the frequency tuning system. It requires much more silicon area and consumes much more power than the frequency controller. Moreover, the precision of the BW loop is sensitive to the offset voltages and so its accuracy is in the order of 5 to 15 %. The first question is: the accuracy of the BW system being limited, is it possible to reduce the complexity of the loop and to increase the accuracy of the loop? The main problem of this system, as in the voltage controlled filter based system, is that the loop control signal is the signal that we want to measure. The accuracy of the system depends of the matching of the envelopes. However, the error voltage is the difference of the envelopes. This is a low-frequency signal and is thus affected by the dc offset voltages of the OTA. The design of a less sensitive tuning loop should require a different architecture. This design should use an indirect tuning method, as this used in the frequency controller. Another quite important question to be answered is: is this accuracy of the loop sufficient enough for the design of continuous-time filters? The answer of this question requires a detailed discussion.

The transfer function of a fourth-order band-pass filter. In filter theory, it has been known for a long time that the transfer function of a biquadratic filter is more sensitive to ω_0 variations than to Q variations [4, 5]. Moreover, the maximum of these sensitivities are in the order of Q and 1, respectively. In biquadratic filters, these maximum sensitivities occur at the -3 dB frequencies for ω_0 and at $\omega=\omega_0$ for Q. Thus, the most important effects of these errors are reflected in the pass-band of the filter. The effects of ω_0 and Q variations on the transfer function of a fourth-order band-pass filter are depicted in Fig 4.18. The ideal center frequency and bandwidth of the filter are 10.7 MHz and 300 kHz, respectively. The pass-band ripple of the filter is 0.2 dB and the quality factor for both biquadratic sections is 20. In the top traces (5 curves) of this figure, the solid trace corresponds to the ideal transfer function. The other 4 curves correspond to the same transfer function but with Q errors in the biquadratic sections of 5%. In these curves, when both Qs change in the same direction, either both Qs increase or both Qs decrease, the pass-band gain is affected by around 0.5 dB. If both Qs increase then the pass-band gain increases but the bandwidth and the ripple of the filter are little affected. The shape of the filter is also little affected when both Qs decrease. The reason is because the Q sensitivity of the biquad is positive. Thus, the Q error of each biquad produces variations on the filter transfer function around the resonant frequency of the biquads. Those variations are proportional to the Q variations. When the Q errors in both biquadratic sections are with the same sign then the transfer function is affected almost in the same way in the whole pass-band range. The situation is different when the Q errors are with a different sign, e.g. the Q of 1 biquad increases and the Q of the other one decreases. In this case, one side of the pass-band gain increases and the other side decreases. This effect is evident in the Fig. 4.18. The Q errors with different sign are the result of mismatches between the biquadratic sections. On the other hand, Q errors with the same sign mean errors in the tuning system. Hence, for narrow band filters, the Q mismatches in the sections are more critical than the Q tuning errors, provided the tuning errors are small e.g. less than 10 %.

Because the fourth-order filter is composed of 2 biquadratic sections, there are 4 possible combinations of ω_0 errors. A possibility is shown in the bottom traces of Fig. 4.18. In the bottom traces, the 3 curves correspond to the magnitude of the transfer function with Q errors of 0%, +5%, -5%, respectively. The errors in ω_{02} and ω_{01} ($\omega_{02} < \omega_{01}$) are -0.5% and +0.5%, respectively. From this curves, it is clear that the transfer function of the fourth order band-pass filter is quite sensitive to the ω_0 errors. In this case, the whole filter parameters are affected. Although the small ω_0 mismatches, their effects on the transfer function are evident. A pass-band gain attenuation by more than 2 dB, the ripple of the filter increases from 0.2 dB up to 0.6 dB and the bandwidth changes by around 50 %. Again, the Q errors, with $\Delta Q_1 = \Delta Q_2$, have little effect on the frequency response of the filter.

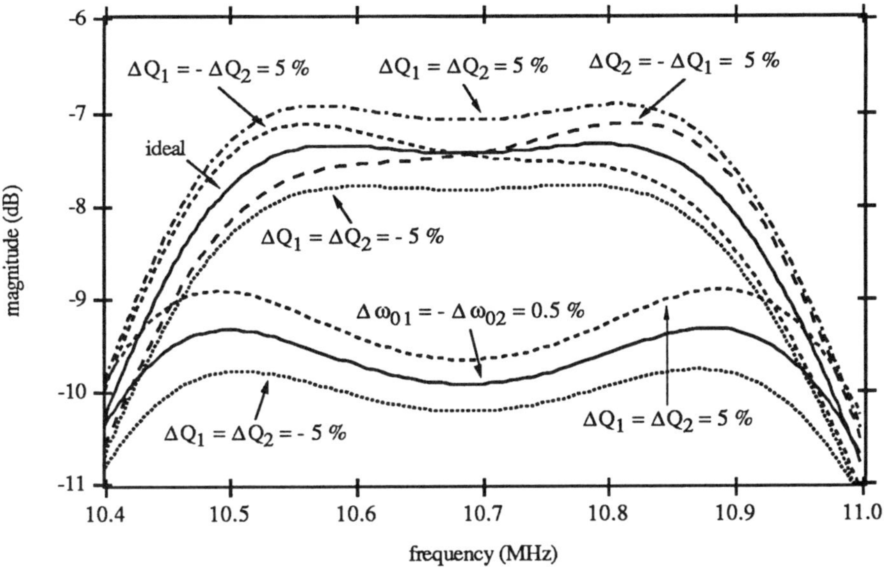

Fig. 4.18. Effects of the ω_0 and BW mismatches. In the top traces the ω_0 errors are zero. In the bottom traces, 0.5 % variation on ω_0 and 5 % Q errors in the biquadratic sections are considered.

If the ω_0 errors of the biquadratic sections have the same sign then the center frequency of the filter shifts to either higher frequencies or lower frequencies depending of the sign of the ω_0 errors. Also, if the ω_0 errors of the biquads are the opposite to the errors considered in Fig. 4.18, then the opposite effects appear in the transfer function i.e. the pass-band gain increases and the bandwidth decreases.

It is important to recognize that these effects are smaller for lower Q transfer functions because the ω_0 sensitivity of the biquads is proportional to Q [4, 5]. The conclusions are:

1) The Q errors are not quite important if they occur in the same direction. Hence the Q mismatches between the biquadratic sections are more important than the absolute value of the Qs, provided the Q variations are lower than 10%. Fortunately, this is the case in the proposed technique. In fact, the absolute value of the Qs are not very well controlled, but, because the Q of the sections are in the same order of magnitude, the mismatches between them can be further reduced.

2) According to these results, it is quite more important to control very accurately the frequency of the poles than the exact value of the quality factor, providing that the Q mismatches in the structures are small; e.g. less than 1-2%.

4.6. Conclusions

The major limitations of the continuous-time filters have been reviewed. The advantages and drawbacks of two different implementations of second-order filters have been discussed. The optimum dynamic range and lower pass-band sensitivity of the ladder filters makes them very attractive for broad-band applications. For the biquad, the dynamic range of the low-pass filter is very bad. However, the higher programmability of the biquad and its good dynamic range for band-pass filters makes this option attractive for the implementation of narrow band filters. With respect to the automatic tuning of the filters, the biquads can be tuned more accurately than the ladder filters.

The major disadvantage of the continuous-time filters is the unavoidable necessity of additional tuning. For on-chip tuning, a self controller has to be included in the system. The mostly used techniques for this purpose have been reviewed. The advantages and limitations of the phase locked loop based techniques have been pointed out. A major limitation of these techniques is the increased noise level of the filter.

Another approach for the control of the poles frequency based on charge comparison has been described. This technique carries out the automatic tuning of the filter at a different frequency range than the filter pass-band frequency range. This fact reduces the tuning signal feedthrough and hence the dynamic range of the filter is little sensitive to the operation of the automatic tuning system. The precision of the frequency tuning loop, if properly designed, is as good as with the PLL based techniques. The topology is a classical second-order system and its stability can be easily guaranteed for a wide tuning range.

For the automatic tuning of the filter bandwidth, the pulse response of the filter is exploited. In general, the precision in the control of the absolute value of Q of the proposed BW tuning system is not as good as the precision of the frequency controller. Nevertheless, this technique allows to reduce the Q errors from 100-400 % down to 10-30 %. Furthermore, it has been shown that the effect of those errors can be minimized designing the filter with small Q spreads.

REFERENCES

[1] F. E. J. Girling and E. F. Good, " The leapfrog or Active-Ladder Synthesis," Wireless World, Vol. 76, pp. 341-345, July 1970.

[2] J. Orchard, " Inductorless Filters," IEE Electronics letters, Vol. 2, pp. 155, June 1966.

[3] L.T. Bruton, " Multiple-Amplifier RC-Active Filter Design with Emphasis on GIC Realizations," IEEE Trans. on Circuits and Systems, Vol. CAS-25, pp. 830-845, Oct. 1978.

[4] L.T. Bruton, " Sensitivity Comparison of High-Q Second-Order Active Filter Synthesis Techniques," IEEE Trans. on Circuits and Systems, Vol. CAS-22, pp. 32-38, Jan. 1975.

[5] A. S. Sedra and P. O. Brackett, *"Filter Theory and Design: Active and Passive"* , Matrix publishers, Inc., Champaign Illinois, 1978.

[6] H. Nevarez-Lozano, A. Hill and E. Sanchez-Sinencio," Frequency Limitations of Continuous-Time OTA-C Filters," IEEE ISCAS-88, Espoo Finland, pp. 2169-2172, June 1988.

[7] F. M. Gardner, *"Phaselocked Techniques"* , Second edition, John-Willey & Sons, USA, 1979.

[8] K. Radhakrishna, V. Sethuraman and P.K. Neelakantaan, "A Novel Follow the Master Filter," Proc. of the IEEE, Vol. 65, pp. 1725-1726, Dec. 1977.

[9] K. Radhakrishna and G. Venugopal, "A Novel Technique for On-Chip Tuning of Monolithic Filters," Proc. of the IEEE, Vol. 75, pp. 257-258, Feb. 1987.

[10] K. Tan and P. Gray, "Fully Integrated Analog Filters Using Bipolar-JFET Technology," IEEE J. of solid-State Circuits, Vol. SC-13, pp. 814-821, Dec. 1978.

[11] K. Moulding, J. Quartly, P. Rankin, R. Thompson and G. Wilson, "Gyrator Video Filter IC with Automatic Tuning," IEEE J. of Solid-State Circuits, Vol. SC-15, pp. 963-967, Dec. 1980.

[12] M. Banu and Y. Tsividis, "An Elliptic Continuous-Time CMOS Filter with On-Chip Automatic Tuning," IEEE J. of Solid-State Circuits, Vol. SC-20, pp. 1114-1121, Dec. 1985.

[13] H. Khorramabadi and P. R. Gray," High-frequency CMOS Continuous-Time Filters," IEEE J. of Solid-State Circuits, Vol. SC-19, pp. 939-948, Dec. 1984.

[14] C. S. Park and R. Schaumann," Design of a 4 MHz Analog Integrated CMOS Transconductance-C Bandpass Filter," IEEE J. of Solid-State Circuits, Vol. SC-23, pp. 750-758, Aug. 1988.

[15] R. Schaumann and M. A. Tan," The problem of On-Chip Automatic Tuning in Continuous-Time Integrated Filters," IEEE Proc. ISCAS-89, Portland Oregon, pp. 106-109, May 1989.

[16] V. Gopinathan, Y. P. Tsividis, R. K. Hester and K. S. Tan," A 5 V, 7th Order Elliptic Analog Filter for Digital Video Applications," IEEE Intl. Solid-State Circuits Conference, San Francisco, pp. 208-209, Feb. 1990.

[17] P. M. Van Peteghem, S. Rujiang," Tuning Strategies in High Frequency Integrated Continuous-Time Filters," IEEE Trans on Circuits and Systems, Vol. CAS-36, pp. 136-139, Jan 1989.

[18] T. L. Brooks and P. M. Van Peteghem," Simultaneous Tuning and Signal Processing in Integrated Continuous-Time Filters: The Correlated Tuning Loop," IEEE Proc. ISCAS-89, Portland Oregon, pp. 651-654, May 1989.

[19] K. A. Kozma, D. A. Jones and A. S. Sedra," An Adaptive Tuning for Integrated Continuous-Time Filters," IEEE Proc. ISCAS-90, New Orleans, pp. 1163-1166, June 1990.

[20] J. B. Hughes, N. C. Bird and R.S. Soin, "A Novel Digitally Self-Tuned Continuous-Time Filter technique," IEEE Proc. ISCAS-86, San Jose, pp. 1177-1180, may 1986.

[21] Y. Wang, F. Lu and A. Abidi, "A 12.5 MHz CMOS Continuous Time Bandpass Filter," IEEE Intl. Solid-State Circuits Conference, New York, pp. 198-199, Feb. 1989.

[22] J. Silva-Martinez, M. Steyaert and W. Sansen, "A Novel Approach for the Automatic Tuning of Continuous-Time Filters," IEEE Proc. ISCAS-91, Singapore, pp. 1452-1455, June 1991.

[23] M. Steyaert and J. Silva-Martinez, "A 10.7-MHz CMOS OTA-R-C Bandpass Filter with 68 dB Dynamic Range and On-chip Automatic Tuning," IEEE Intl. Solid-State Circuits Conference, San Francisco, pp. 66-67, Feb. 1992.

[24] K. Mossberg and D. Akerberg, "Accurate Trimming of Active RC Filters by Means of Phase Measurements," IEE Electronics letters, Vol. 21, pp. 520-521, Sept. 1969.

[25] Y. Tsividis and P. Antognetti, Editors, *"Design of MOS VLSI Circuits for Telecommunications"* , Prentice-Hall, inc., New Jersey, 1985.

[26] L.O. Chua, C. L. Desoer and E. S. Kuh, *"Linear and Nonlinear Circuits,"* Mc Graw-Hill Book Company, Singapore, 1987.

[27] M. Pelgrom, A. Duinmaijer and A. Welbers, "Matching Properties of MOS transistors," IEEE J. of Solid-State Circuits, Vol. 24, pp. 1433-1439, Oct. 1989.

[28] K. Lakshmikumar, R. Hadaway and M. Copeland, "Characterization and Modelling of Mismatch in MOS Transistors for precision Analog Design," IEEE J. of Solid-state Circuits, Vol. SC-21, pp. 1057-1066, Dec. 1986.

[29] J. B. Shyu, G. C. Temes and F. Krummenacher," Random Error Effects in Matched MOS Capacitors and Current Sources," IEEE J. of Solid-State Circuits, Vol. SC-19, pp. 948-955, Dec. 1984.

[30] H. U. Post and K. Waldschmidt," A High-Speed NMOS A/D Converter with a Current Source Array," IEEE J. on Solid-State Circuits, Vol. SC-15, pp. 295-300, June 1980.

[31] K. Warwick (editor), *"Implementation of Self-Tuning Controllers"*, Peter Peregrinus Ltd., London, 1988.

Further readings.

[32] J. M. Khoury, "A 15 MHz Continuous-Time Bessel Filter for Disk Drives," IEEE Intl. Solid-State Circuits Conference, New York, pp. 134-135, Feb. 1991.

[33] T. Kuan and K. Martin, "An adaptive Analog Continuous-Time CMOS Biquadratic Filter," IEEE J. of Solid-State Circuits, Vol. 26, pp. 859 - 867, June 1991.

[34] Z. Czarnul and Y. Tsividis, "Independent Tuning of Quality Factor and Unity-Gain Frequency in a Transconductance-Capacitance Integrator," IEE Electronics Letters, Vol. 22, pp. 1026 - 1027, Sept. 1986.

Appendix A4.1

The transconductance to charge converters

The basic transconductance to charge transducer is shown in Fig. A4.1a. Its equivalent circuit is shown in Fig. A4.1b. The voltage source I_R/g_m is controlled by both the dc current source I_R and the OTA bias current (see Fig. 4.5). From the equivalent circuit of Fig. A4.1b, the capacitor C_1 is charged to the OTA output voltage with a time constant given by

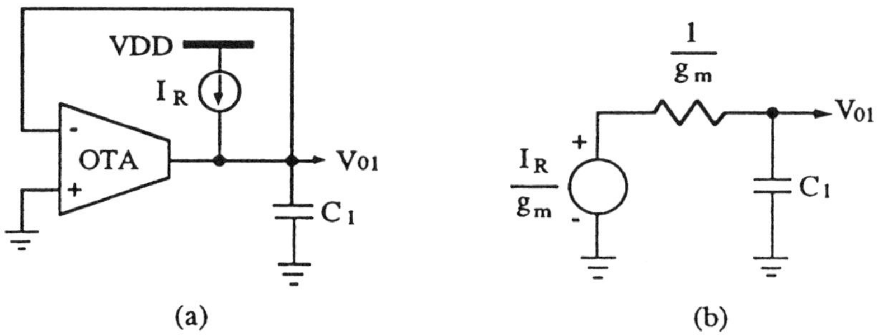

Fig. A4.1. The basic transconductance to charge transducer. a) the block diagram and b) the equivalent circuit.

$$\tau_C = \frac{C_1}{g_m} \qquad (A4.1)$$

For a complete charge of C_1 the time duration of the clock phase ϕ_1 have to be larger than 4 τ_C. This fact limits the clock frequency to values lower than $g_m/8C_1$. In order to decrease the output impedance of the OTA, a buffered structure can be used. A low output impedance structure using a PMOS transistor is shown in Fig. A4.2a. In this topology, the OTA output current is equal to the reference current, I_R. For the magnitude of V_{01}, this is exactly the same situation that for the original circuit (see Fig. A4.1). However, in this circuit the capacitor C_1 is charged to the voltage V_{01} by an equivalent impedance equal to the transconductance of the PMOS transistor. Obviously, this transconductance can be easily designed higher than the OTA transconductance. Hence, the charging time of the transconductance to charge transducer is further reduced. As a result, higher clock frequencies can be used.

A solution using a PMOS-NMOS buffer is shown in Fig. A4.2b. The higher transconductance of the NMOS transistor decreases the charging time of the topology. This advantage makes this topology attractive for very high-frequency applications.

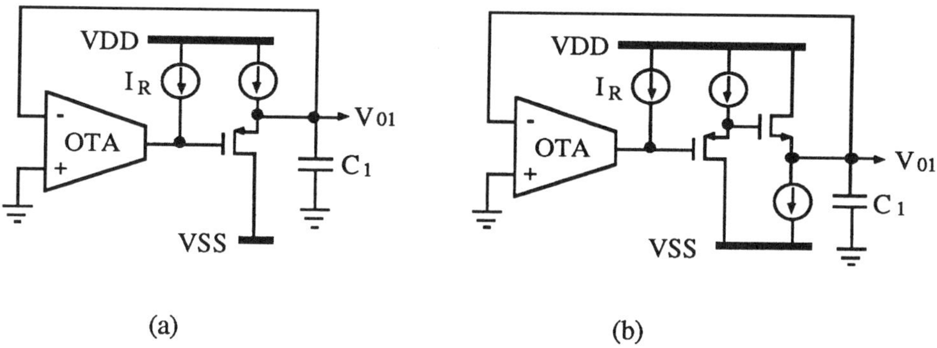

Fig. A4.2. Fast transconductance to voltage converters. a) using a PMOS buffer and b) using a PMOS-NMOS buffer.

CHAPTER

5

HIGH-PERFORMANCE REALIZATIONS OF CONTINUOUS-TIME FILTERS

5.1. Introduction.

In this chapter, several high-performance realizations of continuous-time filters are described. The realizations are based on two of the most popular type of structures, ladder filters and biquadratic based filters [1-15, 17-24]. The frequency of these filters ranges from 100 kHz up to 10.7 MHz. For the design of these filters, the techniques described in chapters 2 to 4 are employed. The employed OTAs use the low-distortion input-stage presented in chapter 2. The quality factor (e.g. the passband ripple) of the filters is controlled by full CMOS resistors. These resistors have been discussed in chapter 3. The automatic tuning systems described in chapter 4 are used for the control of a 10.7 MHz bandpass filter.

In section 5.2, a doubly terminated fourth-order 100 kHz lowpass ladder filter is discussed. For this filter a fully-differential topology has been used. The common-mode feedback technique for fully-differential structures described in the chapters 2 and 3 is employed in this filter and its experimental evidence is added. For the control of the passband ripple, the low-distortion floating resistor based on a transistor biased in triode region is employed. The 100 kHz filter achieves a total harmonic distortion lower than - 60 dB for input signals up to 2 V_{pp}. The intermodulation distortion is lower than - 60 dB for two mixed signals with amplitudes of 1 V_{pp}. For these harmonic distortion components, the dynamic range of the filter is around 66 dB.

A second-order bandpass filter with more than 1 decade programmable Q is described in section 5.3. The center frequency of the filter is 1.8 MHz. The quality factor of the filter is controlled with the current source based resistor described in chapter 3, section 3.5. The extremely large tuning range of this type of resistors is demonstrated in this filter. The Q of the filter is electrically programmable from 2 up to 50. For this Q range, the resistor requires a bias current lower than 25 % of the LDOTA bias current. The experimental results for the filter have shown THD < - 60 dB for input signals up to 1.2 V_{pp}. However, the linear range is limited by the common-mode range of the LDOTA more than by the linearity of the resistor because the topology is single-ended.

The most general design is a fourth-order 10.7 MHz bandpass filter with on-chip automatic tuning system. This filter and the automatic tuning system are described in section 5.4. The major specs for the filter are: center frequency of 10.7 MHz, bandwidth of 300 kHz, passband ripple of 0.5 dB and peak gain of - 4 dB. For this purpose, a fourth-order filter based on 2 biquadratic sections has been implemented. The design considerations for the architecture of the filter, for the OTAs and for the floating resistors are discussed. In addition, the automatic tuning system is included in the chip. For the OTA, a fully-differential LDOTA is used. The ripple and the peak gain of the filter are controlled with two current source based floating resistors. For the control of the filter center frequency, the automatic frequency controller described in chapter 4, section 4.4, is employed. The bandwidth of the filter is adjusted using the transient analysis based technique described in section 4.5.

The experimental results for the maximally flat 10.7 MHz bandpass filter have shown that the center frequency of the filter can be controlled very accurately using the charge based tuning technique. The absolute accuracy of the automatic frequency control system is around 1 %. The frequency controller is able to track the center frequency of the filter to the clock reference from 7 MHz up to 14 MHz. For this design, the accuracy of the filter bandwidth is worse than expected. Instead of a BW error of 15 %, the experimental results showed BW errors of 30 %. However, it has to be remarked that without Q tuning, Q errors of 400 % can be expected. The third-intermodulation distortion of the topology is lower than - 40 dB for a two tone signal with amplitude of 3.2 V_{pp}. Due to this extremely large linear range, the dynamic range of the filter is around 68 dB. The results for the 10.7 MHz bandpass filter are presented in section 5.4. At the end of the chapter, some general comments and also some conclusions are given.

5.2. A High-Performance OTA-R-C Lowpass Filter.

In this section, several of the previously discussed techniques for the design of high-performance fully-differential OTAs are employed. These techniques are used for the design of a fourth-order doubly terminated low-pass filter [14-15]. The filter is based on Operational Transconductance Amplifier-Resistor-Capacitor (OTA-R-C) building blocks. For the voltage-to-current conversion, the fully-differential LDOTA is employed. The design of the LDOTA has already been discussed in section 3.3.2 of chapter 3. The dimensions and the bias conditions of the transistors can be found in table 3.4 of the same section. Hence, the design of the LDOTA is not discussed any further.

For the control of the passband ripple, the low distortion floating resistor based on triode biased transistors is used. The resistor has also been discussed in chapter 3, section 3.5.3. So, in this section, only the filter design considerations are discussed. The technique using active resistors allows to control the ripple of the filter in a very width range without sacrificing the filter dynamic range. Even more, noise, power consumption and silicon area are lower for the resistor than for the commonly used OTA based resistors.

The design techniques are validated by experimental results. The dynamic range of a fourth-order 100 kHz lowpass filter is about 66 dB. For this dynamic range, the total harmonic distortion is lower than - 60 dB. The noise has been integrated from 10 Hz up to 100 kHz. The supply voltages are only ± 2.5 Volts.

5.2.1. The filter architecture. The filter can be designed using either biquads or ladder filters [1 - 7]. The most important characteristics of both approaches have been discussed in the section 4.2 of the previous chapter. In that section, it was shown that the dynamic range of the lowpass ladder filter is superior to the dynamic range of the biquadratic based lowpass filter. Even more, the ladder filter is little sensitive to the tolerances of the components, mainly in the filter passband. The ladder filter is not as versatile as the biquad but it can be tuned. For these reasons, the fourth-order lowpass filter is implemented as a doubly terminated ladder filter. The major specifications for the lowpass filter are shown in table 5.1. The RLC prototype of the fourth-order ladder filter is shown in Fig. 5.1. Based on this prototype and on the specifications, the RLC components can be easily found in already available tables [1] or computed from well known relationships [6].

parameter	specification
filter order	4
cutoff frequency	100 kHz
passband ripple	< 0.5 dB
attenuation @ 600 kHz	> 60 dB
THD	< - 60 dB
dynamic range	> 60 dB
power supply	± 2.5 Volts

Table 5.1. Specifications for the equiripple lowpass filter .

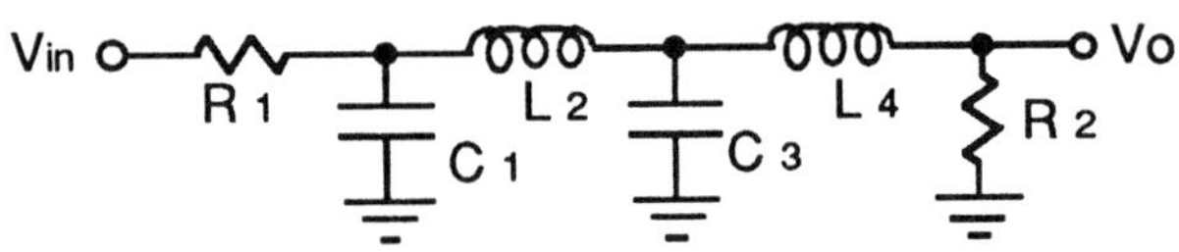

Fig. 5.1. RLC prototype of the fourth-order lowpass filter.

Active realization. For the active realization of the filter, the capacitors are considered as the basic elements. They are naturally implemented in CMOS technology. The resistor can be simulated by the topologies earlier discussed, see chapter 3. For the ladder filter of Fig. 5.1, it is desirable to split the resistor R_1 in a grounded resistor and a transconductor, as shown in Fig. 5.2. The transconductor, g_m, makes the topology more versatile. For the RLC filter, the loops which determine the pole positions are not affected when R_1 is replaced by the "pseudo resistor" of Fig. 5.2. This is because the impedance seen for C_1 and L_2 in the direction of the input signal is R_1. This impedance is the same when the pseudo resistor is employed. The advantage of the pseudo resistor will be evident when the total filter is discussed.

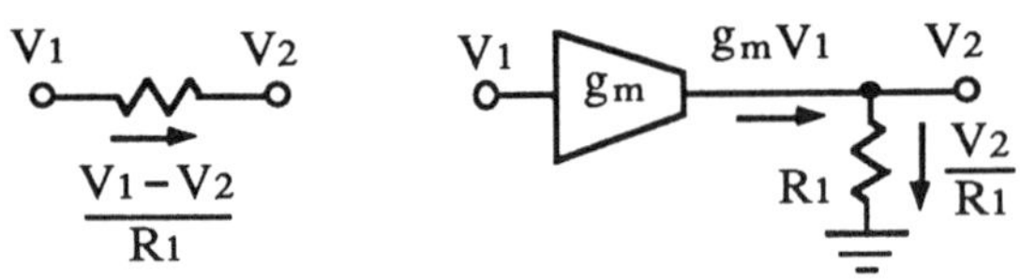

Fig. 5.2. OTA-R implementation of the floating resistor R_1.

The floating inductor can be implemented using feedback and feedforward paths as shown in Fig. 5.3. For this implementation, the equivalent floating inductance is given by

$$L = \frac{C_L}{g_m^2} \tag{5.1}$$

In Fig. 5.3, the non used OTA terminals can be employed for the implementation of additional filter loops. Using these building blocks, the fourth-order lowpass filter can be implemented. The fully-differential version of the doubly terminated lowpass filter is shown in Fig. 5.4.

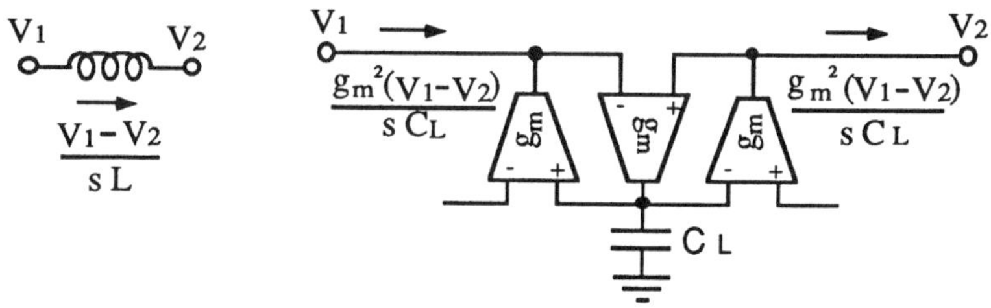

Fig. 5.3. OTA-C implementation of a floating inductor.

The filter includes 7 OTAs, 2 floating resistors and 8 grounded capacitors. The capacitors are grounded because of the single poly technology. The values of the grounded capacitors are given in table 5.2. In the same table, the values of the resistors have been included. The OTA transconductance is 6 μA/V and the bias current I_B is 6 μA. The design of the LDOTA was discussed in section 3.3.2.

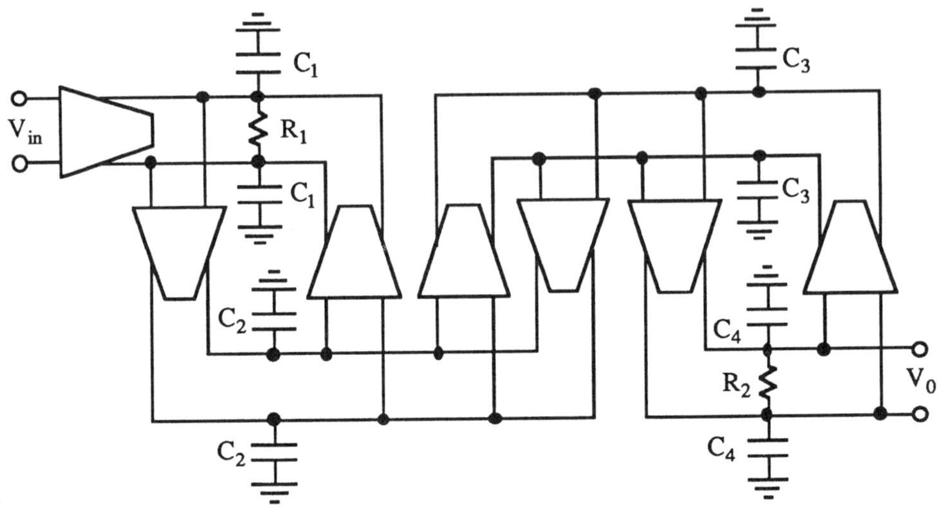

Fig. 5.4. Block diagram of the fourth-order equiripple low-pass filter.

5.2.2. Design considerations. In this design, for the common-mode control of the LDOTAs the techniques discussed in chapter 2, section 2.5.3, are employed. Instead of using one CMFB per OTA, the proposed technique employs one CMFB per filter pole, see Fig. 2.11. As a result, for the designed fourth-order filter, 7 OTAs and only 4 CMFBs are required.

component	value
C_1	9.466 pF
C_2	12.14 pF
C_3	13.88 pF
C_4	8.778 pF
R_1	221.2 kΩ
R_2	272.7 kΩ

Table 5.2. Capacitor and resistor values for the filter.

Form section 3.3.2, the second pole of the common-mode loop was calculated at 1.2 MHz. Under maximum load conditions, the gain-bandwidth product of the common-mode loop is higher than the cutoff frequency of the filter by a factor of 2. Under minimum load conditions, the gain-bandwidth product of the CMFB is in the order of 450 kHz, leading to a phase margin of 65 degrees for the common-mode loop. However, in the filter two output-stages are connected to the common-mode feedback loop. One is used for the direct path and the other one for the feedback. Hence, the gain-bandwidth product of the common-mode loop increases by a factor of 2. An exception is the termination of the filter. However, in this output the lower capacitance has been used. As a result, the gain-bandwidth of the common-mode loop is 400 kHz and 900 kHz for maximum and minimum load capacitance, respectively. The phase margin is around 48 degrees for the minimum loaded CMFB and around 65 degrees for the maximum loaded CMFB. It is quite important to be sure that the phase margin of the common-mode loop is good enough, otherwise common-mode oscillations can appear at the output of the OTAs. These oscillations increase the common-mode settling time of the filter. However, the common-mode oscillations decreases the common-mode range of the filter. In addition, if large mismatches are present, they interfere with the differential signal and so increasing the differential noise of the filter.

The control of the passband ripple is carried out by the linear resistors. The resistor has been implemented using the topology of Fig. 3.12, section 3.5. This approach, instead of the unity feedback OTA resistor, presents larger tunability range without the reduction of the filter dynamic range. In a typical ladder filter, the passband gain is determined by the ratio of the resistors. In this filter, however, due to the use of the pseudo resistor the passband gain is approximately given by

$$A_{V\text{passband}} \approx \frac{g_{m1}}{g_1 + g_2} \tag{5.2}$$

In this case, the passband gain of the filter can be controlled with the transconductance of the first LDOTA of the filter. However, this LDOTA, the first one in Fig. 5.4, does not affect the behavior of the poles. This fact increases the versatility of the filter. For this design, however, all the LDOTAs are equals.

5.2.3. Experimental Results. The fully-differential filter has been fabricated in a standard 1.2 μm single poly N-well CMOS process. A microphotograph of the chip is shown in Fig. 5.5. An additional LDOTA is used to buffer the filter outputs. For several chips, the bias current needed for the adjusting of the filter cutoff frequency was in the range 5.5 - 7.5 μA. The discussed results are for a bias current of 6.5 μA.

Fig. 5.5. Microphotograph of the chip.

Frequency response. The measured magnitude and phase response of the filter are shown in Fig. 5.6. The ripple of the filter is easily tunable within a decade (from 0.2 dB up to 5 dB) with low effect on the THD, as expected. The stop band attenuation is about - 70 dB at 600 kHz. For a fully-differential filter, 4 transfer functions can be defined. The first one is the typical transfer function of the system, shown in Fig. 5.6, defined as the ratio of the differential output to the differential input. The other three transfer functions of the filter are depicted in Fig. 5.7.

The differential-out to common-in transfer function, A_{cd}, determines the common-mode rejection ratio of the filter. The attenuation is of the order of - 70 dB for the whole passband frequency range. In reality, this attenuation is lower but we were unable to measure precisely, mainly due to the setup signal feedthrough. It can be seen that the CMRR (A_{dd}/A_{cd} with A_{dd} the transfer function of Fig. 5.6) is larger than 60 dB for the whole passband frequency range.

The common-mode gain A_{cc} is defined as the common-mode output to common-mode input ratio. According to Fig. 5.7, A_{cc} is lower than - 55 dB in the whole passband frequency range. The differential signal produces a common-mode output voltage as well. The differential in-common-out transfer function, A_{dc}, is in the order of -58 dB for the whole passband frequency range.

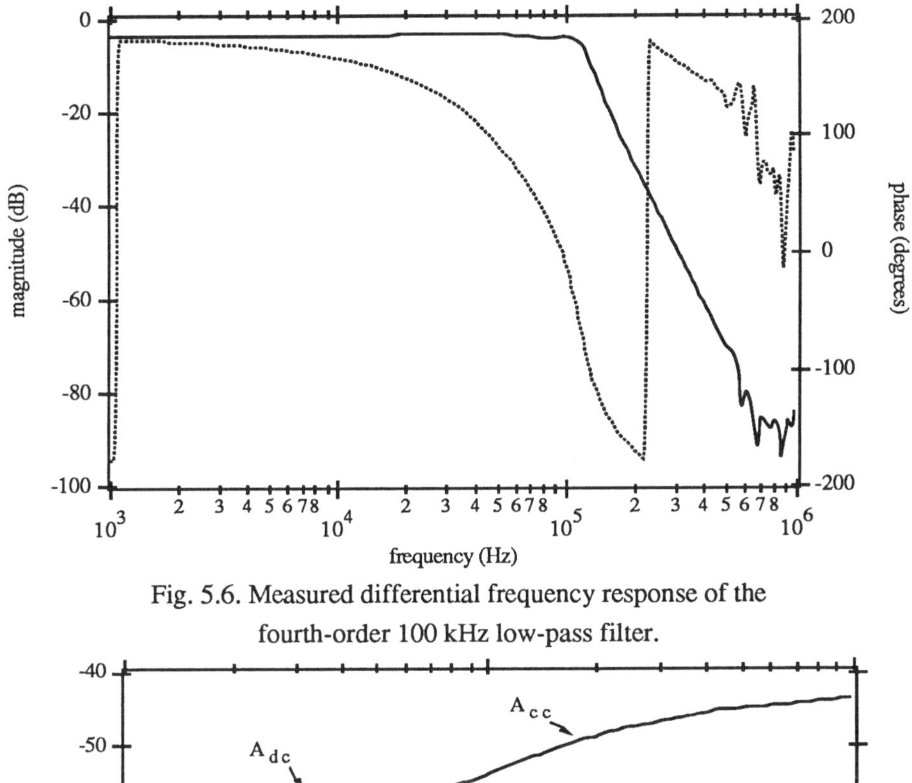

Fig. 5.6. Measured differential frequency response of the fourth-order 100 kHz low-pass filter.

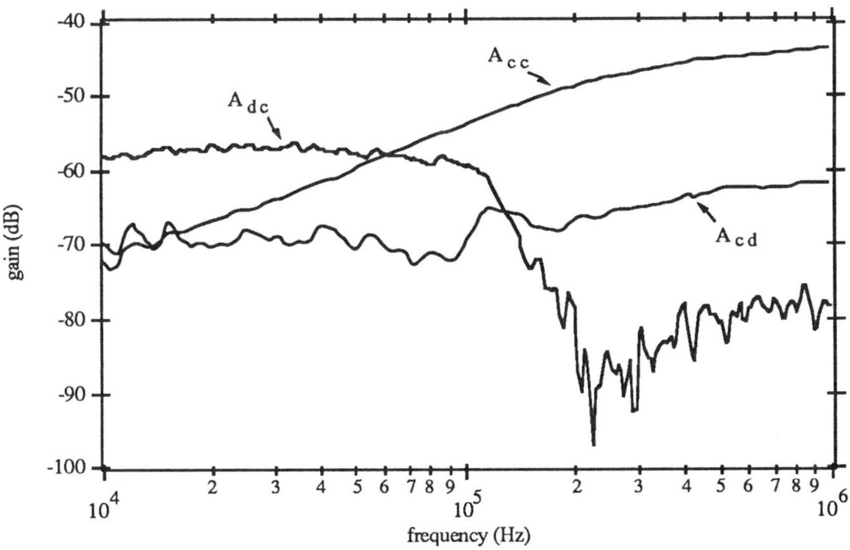

Fig. 5.7. Experimental transfer functions for the fully-differential LP filter.

Rejection to common-mode noise. For the filter, the worst case power supply gains are depicted in Fig. 5.8. In this plot, v_O / v_{dd} is the single-ended v_{dd} voltage gain. Similarly, v_O / v_{ss} is the single-ended v_{ss} voltage gain. The bottom curves are the measured differential gains.

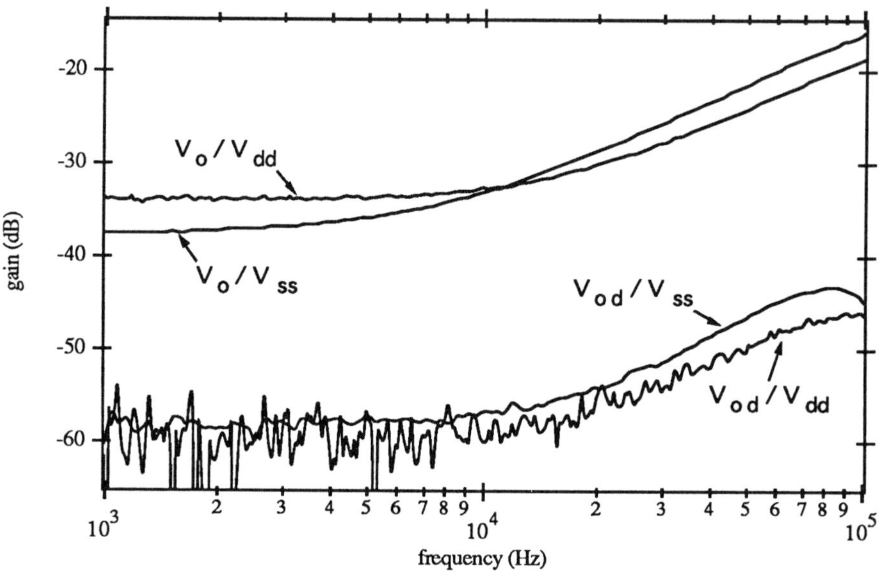

Fig. 5.8. Worst case differential and single-ended power supply gains.

For these results, both the control of the CMOS resistors and the OTAs bias current control are not strongly ac tied to the supply rails. In addition, the effect of the additional OTA used to buffer out the filter signal is present. Hence, these results represent the supply gains for the filter under the worst case. Typically, when the filter control terminals are further ac tied to the supply rails, the supply gains decrease by 4 to 6 dB. According to Fig. 5.8, the differential v_{dd} supply gain is about - 46 dB. For v_{ss}, this gain is around - 43 dB at 100 kHz. The typical results for several chips are summarized at the end of this section.

Harmonic Distortion. The measured HD3 and the total harmonic distortion are shown in Fig.5.9. For these results, a 20 kHz input signal has been employed. It can be noted that these results are quite similar to the OTA distortions, see Fig. 3.4 of chapter 3. The reason is that in the filter the signal is larger only for the first OTA. For the other OTAs, the signal is attenuated almost by a factor of 2. This is due to the natural voltage divider of the lowpass ladder filters [1, 3, 6,7]. Hence, the distortion is mainly due to the first OTA. In these results, the THD is lower than -58 dB for input signals up to 2 V_{pp}. These results also demonstrate the very low distortion predicted for the floating resistor. It can be noted that the harmonic distortion increases rapidly for voltages higher than 2 V_{pp}. The major reason of this is the limited common-mode range of the OTA input-stage. For this design, the common-mode range of the OTA is in the order of 0.7 Volts. In addition, the linear range of the voltage-to-current transducer is limited to input signals smaller than 2 (V_{GS1} - V_T), see section 3.2.3.

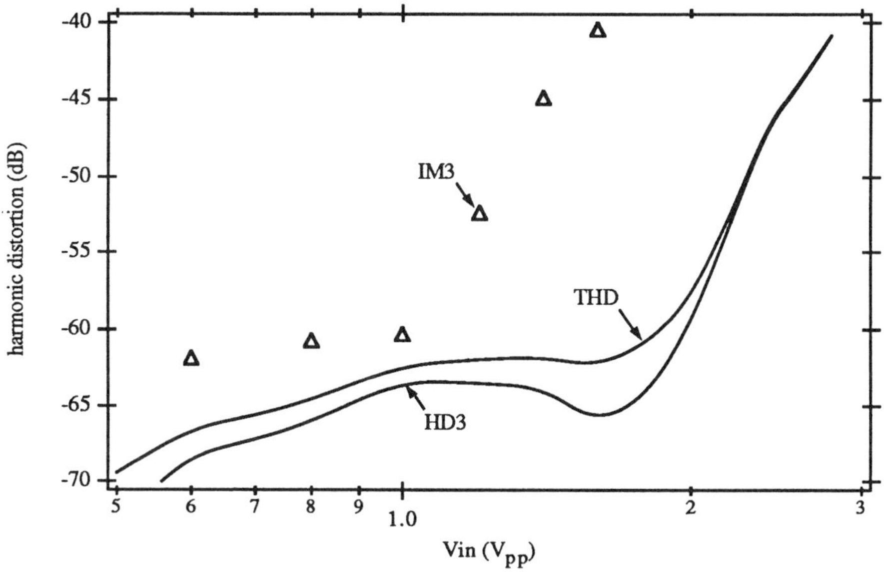

Fig. 5.9. Measured harmonic distortion components for the filter.

In the same plot, the third-intermodulation distortion is shown. For this measurement, 2 equal magnitude input signals have been mixed. For IM3, the abscissa axis is the amplitude of each mixed signal. Thus, the actual peak to peak amplitude of the mixed signal is twice the value associated with the abscissa axis. According to Fig. 5.9, the IM3 is about - 60 dB for input signals as large as 1 V_{pp}.

Dynamic Range. The output referred noise density for the filter and for the OTA and the background noise density are depicted in Fig. 5.10. For the OTA, a load impedance consisting of a floating resistor of 150 kΩ and 2 grounded 10 pF capacitors have been used (see section 3.3.2 for setup details). The integrated input referred noise of the "OTA" integrated from 10 Hz up to 100 kHz is around 150 μV_{RMS}. In these measurements the noise of the setup and the OTA used for the common-mode feedback are included. In addition, a 50 Hz component and its harmonic components are also present. The input referred noise for the single OTA is calculated to be in the order of 90 μV_{RMS}. According to this value, the dynamic range of the OTA is in the order of 79 dB. This value is in good agreement with the 78 dB expected from the calculations carried out in section 3.3.

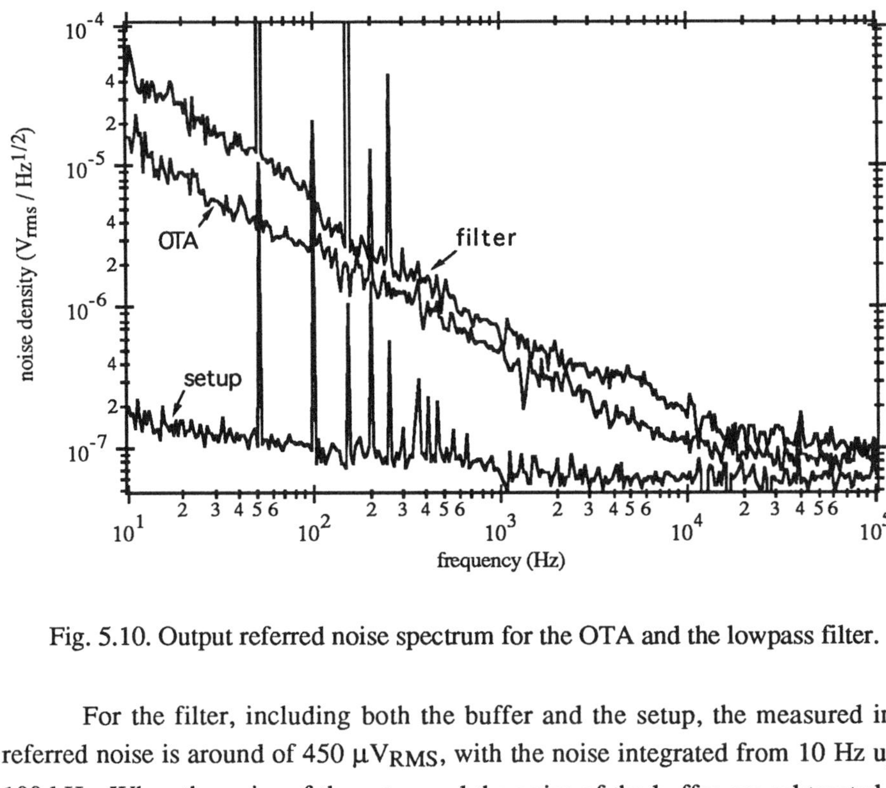

Fig. 5.10. Output referred noise spectrum for the OTA and the lowpass filter.

For the filter, including both the buffer and the setup, the measured input referred noise is around of 450 μV_{RMS}, with the noise integrated from 10 Hz up to 100 kHz. When the noise of the setup and the noise of the buffer are subtracted, the input referred noise of the filter is about 350 μV_{RMS}. Because the harmonic distortion of the filter is lower than -60 dB for input signals up to 2 V_{pp} (0.7 V_{RMS}), this leads to a filter dynamic range of approximately 66 dB.

5.2.4. Summary of results and conclusions. Additional experimental results are shown in table 5.3. The third-intermodulation distortion is lower than -60 dB for two mixed signals of 1 V_{pp}, the total amplitude of the mixed signal is equal to 2

V_{pp}. The filter dynamic range with THD < -60 dB is about 66 dB. The dynamic range can be increased but both power consumption and silicon area increase too. There is a trade off among these parameters.

Typically, the VDD and VSS power supply rejection ratios are in the order of 50 dB and 45 dB, respectively. The rejection to supply noise is expected to increase by improving the symmetry of the layout. Due to the mismatches in the detection of the common-mode signal, a 25 to 30 mV common-mode offset voltage is present. The typical power consumption of the filter is 1.9 mWatts. In table 5.4, the active RC, switched-capacitor and OTA-R-C lowpass filter realizations are compared.

parameter	measured
THD @ $V_{in} < 2\ V_{pp}$	< - 60 dB
IM3 @ $V_{in1} = V_{in2} < 1\ V_{pp}$	< - 60 dB
dynamic range @ THD < -60 dB	≈ 66 dB
CMRR @ 100 kHz	63 dB
PSRR, VDD @ 100 kHz	50 dB
PSRR, VSS @ 100 kHz	45 dB
common-mode offset	20 - 35 mV
supply voltages	± 2.5 Volts
power consumption	1.6 - 2.2 mWatts *
chip area	1.1 mm x 0.84 mm

* several chips measured.

Table 5.3. Fourth order lowpass filter, typical experimental results.

parameter	Active RC	SC	OTA-R-C
f_0	150 kHz	90 kHz	100 kHz
f_0 variations	± 50 kHz	± 1 kHz	± 30 (± 1) kHz *
THD @ 2 V_{pp}	< -80 dB	< -70 dB	< -60 dB
In-band noise	43 μV_{RMS}	280 μV_{RMS}	350 μV_{RMS}
Current drain	800 μA	2800 μA	400 (600) μA *
Chip area	0.4 mm^2	1 mm^2	0.92 (1.5) mm^2 *
Noise $\sqrt{I}$	1.2 $\mu V_{RMS}/\sqrt{A}$	15 $\mu V_{RMS}/\sqrt{A}$	7 $\mu V_{RMS}/\sqrt{A}$

*the data in the brackets include the automatic tuning system.

Table 5.4. Experimental results for several IC active filter realizations.

The active RC filter has the lowest harmonic distortion, the lowest noise level and the minimum chip area. However, this realization presents the largest cutoff frequency variations. Due to the lack of tunability of the resistors and the capacitors it is very difficult to tune this topology. For this purpose, capacitor banks have to be used and so increasing both complexity and silicon area. The switched-capacitor realization presents the highest precision but its power consumption is the highest one. The total harmonic distortion is in between the active RC and the OTA-R-C filters. The OTA-R-C realization presents the highest harmonic distortion, around -60 dB, and the highest noise level. However, the power consumption of this realization is the smallest one.

5.3. A 1.8 MHz second-order bandpass filter with programmable Q over more than 1 decade.

The fourth-order lowpass filter presented in the previous section is a low-frequency and low-Q filter. In this section, a high-frequency bandpass filter is introduced. Since the active filters are very sensitive to the excess phase of the active blocks, the phase response of the building blocks is of primary importance in the design of high-frequency filters. In narrow band filters, the time constant associated with the bandwidth is larger, at least Q times, than the time constant of the resonant frequency. The problems associated with the design of high-frequency narrow band filters are considered. As an example, a bandpass filter with large programmable quality factor is described. In the programmable-Q filter, the control of the center frequency is carried out with two 150 μA/V single-ended LDOTAs. For the control of the Q the current source based technique is used (see section 3.5.4). The Q can be controlled from 2 up to 50. This technique avoids the existence of additional parasitic poles. Moreover, most of the distortion introduced by this simulated resistor is of even order. It can thus be further reduced by current or voltage subtraction.

5.3.1. The programmable-Q bandpass filter. The major goals of this filter are twofold: firstly to allow the phase characterization of the LDOTA ; secondly, to demonstrate experimentally the predicted characteristics for the current source based resistors. For these purposes and due to the higher programmability of the biquad, this approach has been chosen. The design of high Q bandpass filters requires an OTA at the input of the filter with lower transconductance. In the next section a 10.7 MHz filter is described in which the OTA at the input of the filter is Q times lower than the transconductance of the OTAs employed in the loops. In order to avoid the effects of the lower transconductance OTA a special type of bandpass filter is employed. It is recognized that this topology presents several practical limitations. However, it is used only for the characterization of the proposed techniques.

The filter topology. The design of the LDOTA has been discussed in detail in chapter 3, section 3.3. In this section, however, the single-ended LDOTA is not discussed any further. The center frequency of the filter is controlled by a classical second-order loop. For the high Q filters, a loop with large time constant (Q times larger than the time constant associated with the second-order loop) is needed. The implementation of this loop using unity feedback connected OTAs is impractical because its frequency response degrades for smaller transconductances [5,9-13]. The use of partial positive feedback allows to use transconductances in the same range but the sensitivity to component variations increases further [18]. Another approach uses triode biased transistors. However, the frequency response of these simulated resistors is also limited, mainly for large time constants for which long transistors are needed.

In this design, for the control of the filter quality factor the current source based resistor is employed. This type of resistors has been discussed in section 3.5.4. For the filter topology, a special type of "bandpass filter" is employed. The filter is shown in Fig. 5.11.

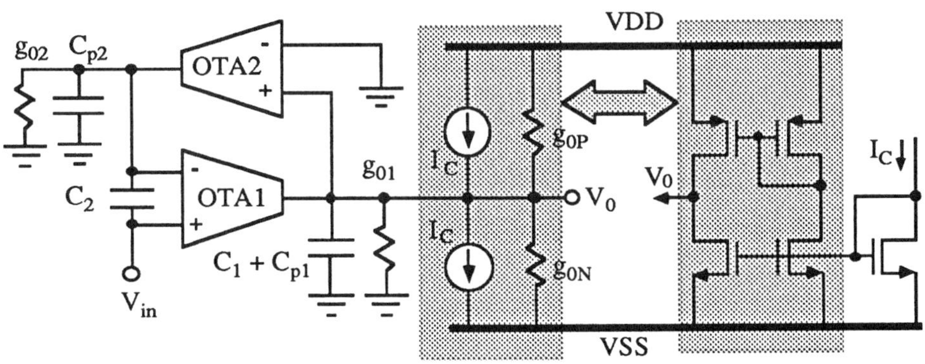

Fig. 5.11. The programmable Q bandpass filter.

In this topology, g_{0i} and C_{pi} are the OTA output conductance and the OTA output parasitic capacitance, respectively. The frequency of the poles is determined by the ratio of the transconductors to the load capacitors. The bandwidth of the filter is determined by the output impedances of the current sources and by the OTA excess phase shift. Detailed expressions for the resonant frequency and filter bandwidth can be found in section 4.2.

At very low frequencies the effects of the capacitors can be neglected. Hence, the topology becomes an amplifier with a load impedance determined by the active resistor. In addition, the OTA2 is connected in a feedback configuration. For this topology, the low-frequency gain of the feedback connected OTA2 determines the low-frequency gain of the filter. Hence, the low-frequency gain of the filter allows to compute quite accurately the low-frequency behavior of the OTA2. At the resonant frequency, however, the gain of the filter is not very well controlled. This is because at the negative input of the OTA1 there is a high pass effect (due to C_2, C_{P2} and g_{02}). Hence, the peak gain of the filter depends strongly on parasitic capacitor C_{P2} and g_{02}. Nevertheless, the voltage divider at the input of OTA1 allows the characterization of high Qs without premature saturation in OTA1. The biquadratic filter is not exactly a bandpass filter since the conductance g_{02} creates a finite transmission zero at the frequency g_{02}/C_{P2}. In practice the capacitor C_{P2} includes all the parasitic capacitors connected to the OTA2 output node.

Neglecting the non-linearities of which the most important component is the second one, the output conductance of the transistors is proportional to the bias current and inversely proportional to the transistor length. In a standard CMOS 3 μm N-well process the Early voltage V_{EN} for the NMOS transistors is in the order of 4.5 Volts/μm and twice that for the PMOS transistors. The output admittance of the current source is the parallel connection of the output conductance of the N transistor, of the P transistor and the drain capacitors. These capacitors are not very important because they can be absorbed by the automatic tuning system. The details of the current source based resistors have been discussed before in the sections 3.5.3 and 3.5.4. For hand calculations, the current source output conductance can be computed from

$$g_0 = I_C\left(\frac{1}{L_N V_{EN}} + \frac{1}{L_P V_{EP}}\right) \approx 0.055\ I_C \quad (A/V) \tag{5.3}$$

with L_N and L_P the channel length of the NMOS transistor and the PMOS transistor, respectively. In this equation, $L_N = L_P = 6$ μm has been assumed. The advantages of this technique are evident from this equation. The conductance is proportional to the bias current. Then the resistor can be tuned over a very wide range. Large variations on the bias current of the transistors do not reduce the common-mode range of the filter. This is because the source of the transistors are tied to the supply rails. Hence, the drain voltage of the transistors can be varied over a very wide range. Moreover, in chapter 3 it was shown that the linearity of the transistor conductance is little sensitive to the bias current. Hence, the dynamic range of the filter is not reduced for the large variations of the load conductance, provided that the LDOTAs are operating in their linear region. A detailed analysis of this topology has been carried out in chapter 3, section 3.5.

5.3.2. Experimental results. In order to buffer the filter output, a high-frequency OPAMP connected in unity feedback configuration has been included on the chip. The gain-bandwidth product of the OPAMP is around 200 kHz. A microphotograph of the chip is shown in Fig. 5.12. This chip was fabricated in a 3 μm CMOS process. The process parameters can be found in appendix A1.1. In the chip, the buffer, a low-distortion single-ended OTA and the biquadratic filter have been included. The experimental results for the OTA were discussed in chapter 3. A discussion of the experimental results for the filter follows.

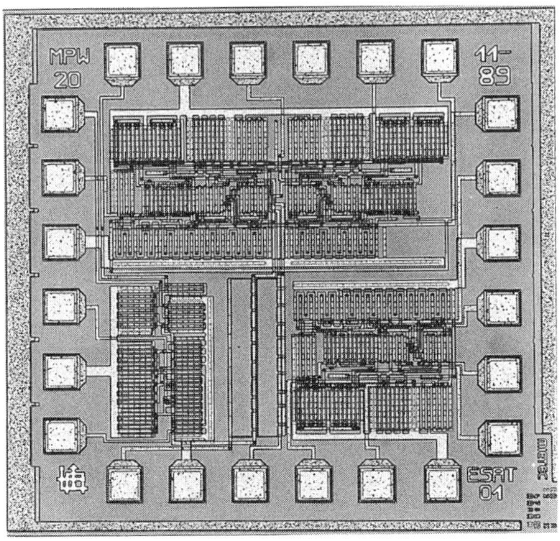

Fig. 5.12. Microphotograph of the chip

Frequency response of the filter. The frequency response of the filter for different bias currents in the active load is shown in Fig. 5.13. The bias current of the current sources is 10 μA (Q ≈ 50), 75 μA (Q ≈ 15), 150 μA (Q ≈ 4) and 250 μA (Q ≈ 2), respectively.

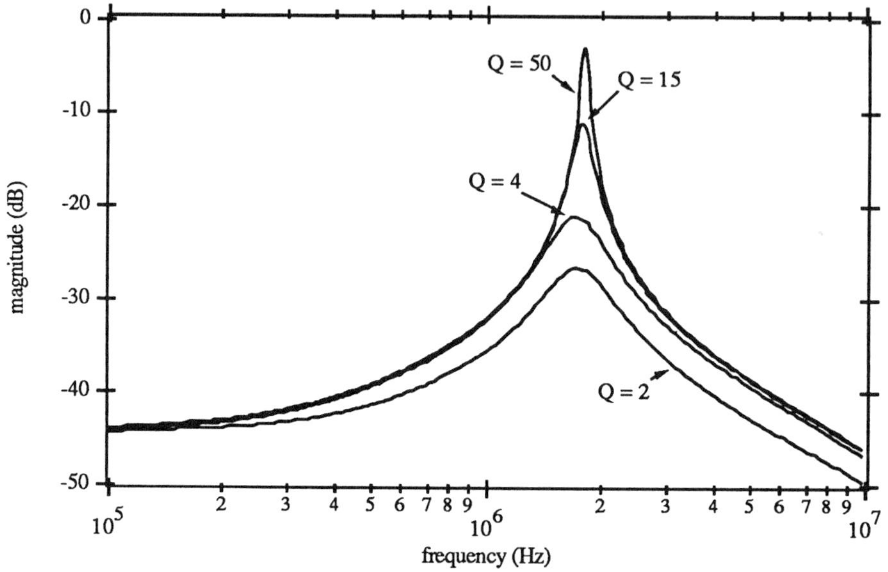

Fig. 5.13. Filter frequency response for different currents in the active load.

The peak gain is, almost, inversely proportional to the filter bandwidth and almost proportional to the Q factor. The wide Q tuning range of this approach can be clearly noted. The quality factor is limited to the order of 50 due to the dominant effect of the zero of the LDOTA. These results demonstrate that the high-frequency response of the filter is dominated by the dominant zero of the LDOTA input-stage, see chapter 2, section 2.3. Even more, the circuit is stable for a load impedance bias current equal to zero. In theory, under this condition the output impedance of the current sources is very high. Hence, the non-zero filter bandwidth is the result of the OTAs excess phase shift. The quality factor of the filter measured under these conditions is around 55. This result implies that the equivalent frequency of this zero is around 100 times higher than the resonant frequency of the filter, see equation 4.16. The low-frequency gain of the filter is about - 45 dB. This attenuation corresponds to a low-frequency gain of the OTA2 of 45 dB.

The phase response of the filter for the same specifications is depicted in Fig. 5.14. The effect of the zero due to g_{02} can be clearly noted in both the magnitude and phase response. In Fig. 5.14, the phase response of the system is degraded for high frequencies, e.g. at frequencies higher than 5 MHz. This is a result of the limited frequency response of the buffer. At high frequencies the finite parameters of the OPAMP lead to an additional finite pole at the output of the buffer.

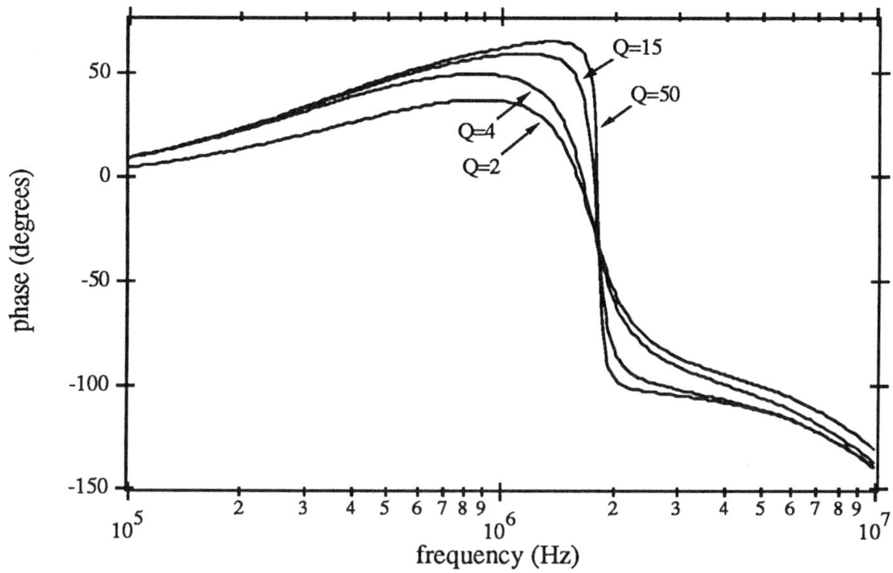

Fig. 5.14. Filter phase response for several resistor bias currents.

Harmonic and intermodulation distortions. The measured harmonic distortions for the filter are depicted in Fig.5.15. The quality factor for these results is 10. In general, the total harmonic distortion for the filter is lower than -60 dB for input signals up to 1 V_{pp}. It can be noted that for voltages larger than 1.2 V_{pp} the harmonic distortions increase rapidly. This is partially due to the voltage gain at the output of OTA2. Moreover, this voltage is in phase with the input signal but with larger swing producing premature saturation at the output of OTA2. The differential signal, however, is of the same order as the input voltage. The other major reason of the limited linear range of the filter is the limited common-mode range of the LDOTA. it has to be remembered that this topology is a single-ended one.

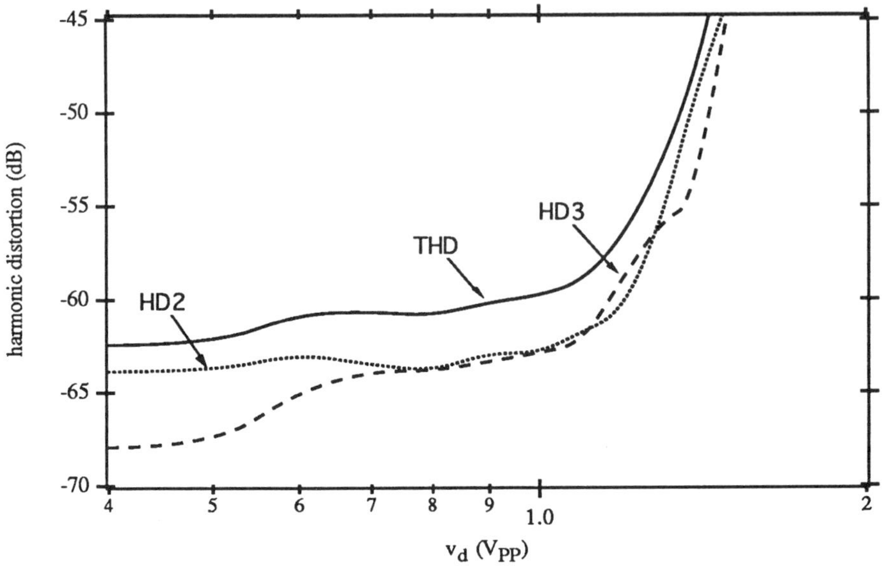

Fig. 5.15. Measured harmonic distortions and THD for the filter.
The quality factor is 10.

Ideally, at the resonant frequency of the filter, the behavior of the bandpass filter is dominated by the active resistor and the excess phase shift of the LDOTA. For moderate Qs, the effects of the resistor should be dominant. However, due to the used topology several parasitic effects are present. For the LDOTA1 most of the signal is common-mode signal. Hence, the peak gain of the filter is affected by the common-mode non-linearities of LDOTA1. Taking into account the finite common-mode gain of the LDOTAs, it is not difficult to demonstrate that the filter transfer function becomes

$$H(s) \approx \frac{g_{m1} C_{P2}}{(C_1 + C_{P1})(C_2 + C_{P2})} \frac{s\left(1 + \frac{2 C_2}{C_{P2}} \frac{g_{mc1}}{g_{m1}}\right)}{s^2 + s \frac{g_{01}}{C_1 + C_{P1}} + \frac{(g_{m2} + g_{mc2})(g_{m1} - g_{mc1})}{(C_1 + C_{P1})(C_2 + C_{P2})}} \tag{5.4}$$

with g_{mci} the common-mode gain of the ith LDOTA. In this expression, the finite output conductance of the LDOTA has been neglected. For the LDOTA, the common-mode rejection ratio is about 40 dB. Hence, the effects of the common-mode gain on the poles can be neglected. Nevertheless, for the peak gain the common-mode gain of LDOTA1 have an enormous effect. For this design, the ratio C_2/C_{P2} has been calculated around 30-50. Hence the coefficient of s in 5.4 is further affected by the common-mode gain of LDOTA1. The physical reason of this effect of the common-mode gain on the transfer function is because the input signal of LDOTA1 is mostly common-mode signal. Hence, the non-linearities of the common-mode gain introduces additional harmonic distortion in the filter response. However, the peak gain of the filter is further affected by the parasitic capacitance C_{P2}. Obviously, the non-linearities present in this capacitor have a direct effect on the harmonic distortion components of the filter.

The third-intermodulation distortion measured at the center frequency of the filter is shown in Fig. 5.16. For Q=15, IM3 is lower than -50 dB for input signals smaller than 0.6 V_{pp}. According to the frequency response of the filter, Fig. 5.11, the peak gain of the filter is about -20 dB. This means that the IM3 is smaller than -50 dB for output signals smaller than 120 mV_{pp}. IM3 increases drastically for input signals higher than 0.8 V_{pp} and for Q=30 this threshold moves down to around 0.4 V_{pp}. The increment of the intermodulation distortion is mainly due to the increment of the signal level at the output of LDOTA2. The peak gain at the output of LDOTA2 is higher for higher Qs indeed.

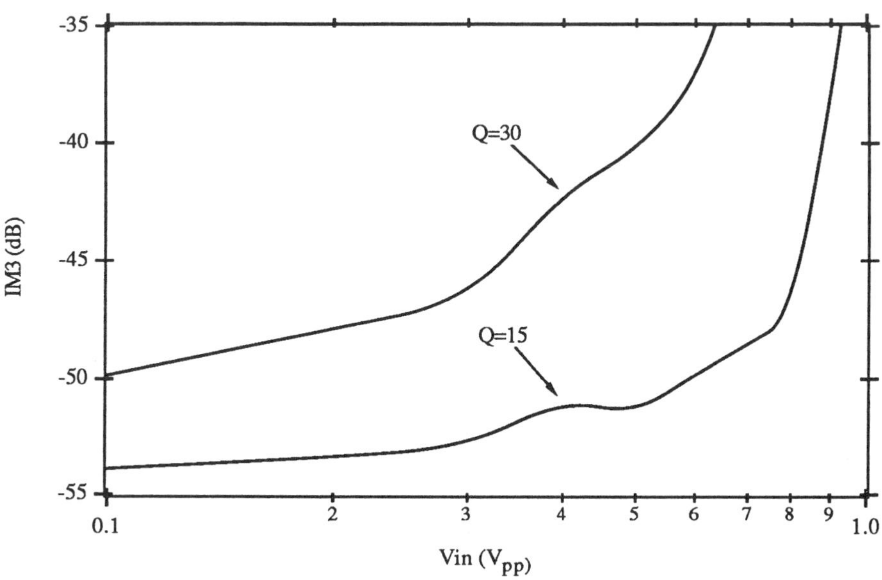

Fig. 5.16. Measured third-intermodulation distortion for the filter.

5.3.3. Further discussion of the results. In this section, the major conclusions for this design are given.

The frequency behavior of the LDOTA (g_m = 150 μA/V) has been characterized with the 1.8 MHz bandpass filter. It has been shown that this topology can be used efficiently in the MHz frequency range as well. The high-frequency behavior of the LDOTA is dominated by the zero of the voltage-to-current transducer. For this design, however, the equivalent frequency of the zero is in the range of 100 MHz. Due to this fact, quality factors of the order of maximum 55 have been achieved.

The large tunability range of the source based resistor has been demonstrated. The quality factor of the filter can be programmed over more than 1 decade, e.g. from 2 up to 50. Even more, the needed current for these variations is in the range of 10 μA to 250 μA. The current consumption of a single LDOTA is around 1.5 mA. Hence, the power consumption of the current source based resistor is much lower than the LDOTA power consumption. However, it has to be recognized that the frequency response of the filter degrades when the bias current of the active load is larger than 300 μA. The reason is due to the mismatches in the NMOS and PMOS current sources. Because the large current, the output-stage of the LDOTA is not able to compensate those large current offsets. This topology is well situated for high-Q applications wherein small bias current are needed.

Certainly, the used filter topology in this implementation is not practical. In the next section, a 10.7 MHz bandpass filter employing a practical topology is discussed. The intention of this design was to characterize the high-frequency behavior of both the LDOTA and the current source based resistor. Unfortunately, with this topology it is not possible to characterize the distortion of the resistor. Some harmonic distortions and intermodulation distortion measurements have been presented. Nevertheless, it is not clear that those results correspond to the behavior of the resistor. The low-distortion behavior of the current source based resistor is demonstrated in the next section.

According to the filter results, it is clear that the frequency response of the LDOTA is quite good. It can be noted in both magnitude and phase response plots that the behavior of the filter is the classical behavior of a biquadratic filter. At high-frequency all the curves, high Q and low Q, are degraded due to the limited frequency response of the buffer. Obviously, this is not an effect of the active resistor because it is present for both high and low Q curves.

5.4. A 10.7 MHz CMOS OTA-R-C bandpass filter with 68 dB dynamic range and on-chip automatic tuning.

The integration in CMOS technologies of narrow-band filters in the MHz range is very attractive for video, IF and other applications. In this frequency range, several switched-capacitor and continuous-time filters have been reported [11-13, 19-23]. Here, a maximally flat 10.7 MHz fourth-order bandpass filter with a dynamic range of 68 dB is described. The bandwidth and the ripple of the filter are 300 kHz and 0.5 dB, respectively. The third-intermodulation distortion is lower than - 40 dB (1%) for a two tone input signal up to 3.6 Volts peak to peak. In addition to the filter, a frequency controller and a bandwidth controller are included. For the frequency controller, the clock frequency can be varied from 300 up to 600 kHz. For this clock frequency range, the center frequency of the filter is controlled from 7 MHz up to 14 MHz. The frequency tuning errors are in the order of 1% and the filter bandwidth varies within 30 %. The supply voltages are only ± 2.5 Volts.

In this section, both the filter and the automatic tuning system are described. The design considerations for the low-distortion and large dynamic range 10.7 MHz filter are discussed in section 5.4.1. The design of the automatic tuning system is described in section 5.4.2. Experimental results for the automatically tuned filter are presented in the next section. The characterization of the automatic tuning system is presented in section 5.4.3. In the last section some conclusions are addressed.

5.4.1. The filter architecture and design considerations. A major decision in the design of the system is the filter architecture: ladder or biquad sections. In section 4.2 a comparison of both techniques has been carried out. It was shown that for the implementation of the Q control loop, the ladder filter requires higher conductance's than the biquad needs. In high-order bandpass ladder filters, however, the resonant loops are interconnected by transconductances Q times lower than the transconductances used in the resonators [22]. These transconductances lead to additional loops which make it more difficult to control accurately the frequency of the poles. In addition, the lower transconductances in the loops increase the excess phase shift of the topology. The biquad overcome those shortcomings but its sensitivity of the filter to variations in the components is slightly higher. Moreover, the higher programmability of the biquad allows to tune the center frequency of the filter independently of the BW tuning. For this design, the biquad based bandpass filter shown in Fig. 5.17 has been chosen.

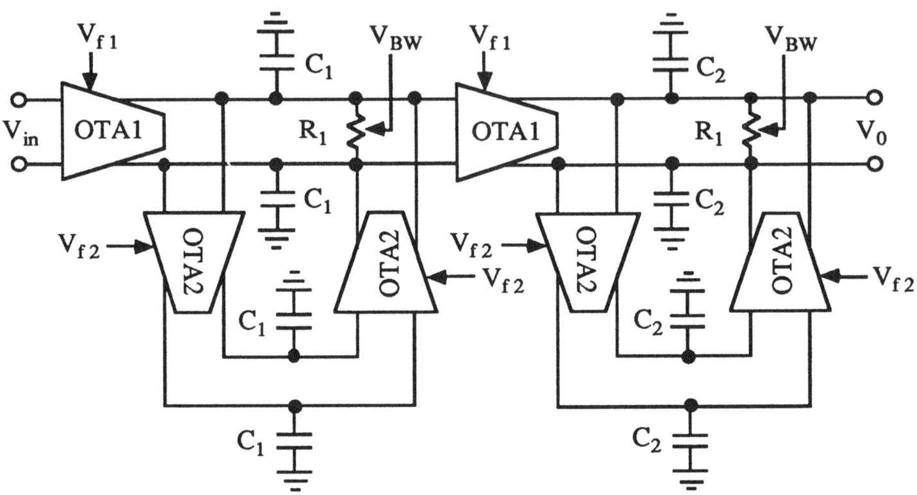

Fig. 5.17. Architecture for the fourth-order maximally flat 10.7 MHz bandpass filter.

The filter architecture is based on 2 biquadratic sections connected in cascade. The pole frequencies, $\omega_0 = 2\pi f_o$, are controlled by the transconductance of the OTA2, g_{m2}, and its associated capacitor $C_{1(2)}$ as $g_{m2}/C_{1(2)}$. Ideally, the quality factor of each biquad is determined by the product $g_{m2}R_1$, see section 4.2. The resistor R_1 is a full CMOS implemented resistor. The peak gain of the biquads are controlled by the transconductance of OTA1, g_{m1}, and the resistor. This gain is given by the product $g_{m1}R_1$. The transconductance g_{m1} is lower than g_{m2} by a factor Q. However, g_{m1} has been designed equal to $1/R_1$. Hence, the peak gain of the second-order bandpass filters is unity. As a result, an almost optimum dynamic range around the filter passband frequency range is obtained. This is because at that frequency range the gain at all the filter terminals are almost unity. A description of the building blocks follows.

The low distortion transconductor. The core of the filter is the voltage-to-current transducer. In order to reduce the even-order harmonic distortions and the common-mode signals, fully-differential LDOTAs are employed. For the capacitors, poly-silicon capacitors of 4 pF have been employed. The total parasitic capacitor at the output of the first biquad is of the order of 0.6 pF. For the second biquad the parasitic capacitor is slightly higher because another transconductor OTA2 has been used to buffer the filter outputs. The total parasitic capacitor at the output of the filter is of the order of 0.9 pF. Hence, the poly-silicon capacitors C_2 have slightly been modified to compensate this effect. The OTA2 transconductance is around 300 μA/V. The OTA1 transconductance is Q times lower, 15 μA/V.

The design procedure for the LDOTA has been discussed in chapters 2 and 3. It is not discussed any further here; only the major design considerations are mentioned. In order to reduce the power consumption of the design, $\beta_1/\beta_2 = 7.5$ and $V_{GS1} - V_T = 0.6$ Volts have been used. HSpice simulations for the filter OTA show THD < -50 dB for input signals up to 3.6 V_{pp}. For the common-mode feedback system, the technique used in the 100 kHz filter is employed in this filter as well. This technique has been discussed in chapter 2, section 2.5. The CMFB systems are shared by several OTAs and so lower power and lower silicon area are achieved. The fourth-order filter uses 6 OTAs and only 4 CMFB systems. The dimensions of the OTA transistors and their bias current are given in table 5.5. In this table, the transistors correspond to the LDOTA of Fig. 2.8, chapter 2. M_6' and M_{fb} are the transistors used in the common-mode feedback system, see Fig. 2.11.

The low distortion full CMOS resistor. In the filter, the theoretical value of the floating resistors is 66 kΩ. Due to the effects of the OTA excess phase shift, it is required to adjust those resistors in a very wide range, e.g. more than 100 %. For this reason, the use of the OTA based resistor is avoided. The active resistors are implemented by the output conductance of two saturated transistors connected in series as shown in Fig. 3.14, chapter 3. The output conductance of the transistors is controlled by their bias currents which are generated by the automatic Q tuning system, V_Q. Since the output conductance of the MOS transistor is almost proportional to its bias current and the quality factor of the biquad is inversely proportional to the resistance, the Q of the filter can be adjusted over a very wide range. The very large tunability of this technique has been demonstrated in the bandpass filter with programmable-Q for more than 1 decade.

transistor	OTA1 W/L (μm/μm)	OTA1 I_{BIAS} (μA)	OTA2 W/L (μm/μm)	OTA2 I_{BIAS} (μA)
M1	12/4.5	16	150/4.5	500
M2	4.5/16	8	20/4.5	0
M3, M4	6/4.5	8	75/4.5	250
M5	100/2.25	56	300/1.5	750
M6	28/4.5	56	200/1.5	750
M6'	-	-	100/1.5	375
M7	12/4.5	8	250/4.5	500
M8	150/2.25	80	300/1.5	1500
M9	32/4.5	8	250/4.5	750
Mfb	-	-	100/1.5	375

Table 5.5. Dimension and bias conditions for the transistors of the LDOTAs used in the 10.7 MHz bandpass filter.

The linearity of the resistor is good enough due to the symmetry of the structure and the low bias current required [25]. More details for the design of the resistor can be found in section 3.5. The theoretical work, the design considerations and some simulations for this resistor can be found in that section as well. For this CMOS resistor, the dimensions and the simulated bias current of the transistors are:

transistor	W/L (μm/μm)	drain current
M_1, M_2	20 / 1.5	20 μA
M_s	20 / 4.5	40 μA

Table 5.6. Dimensions and bias conditions for the transistors of the active resistor (see Fig. 3.13).

5.4.2. The automatic tuning system. For the tuning of the filter, there are two controllers: one for the tuning of the poles frequency and another one for the tuning of the filter bandwidth [24, 26]. The frequency controller adjusts the transconductance of the OTAs through their bias currents. These bias currents are controlled by the voltages V_{f1} and V_{f2} in Fig. 5.17. The frequency controller employs a 450 kHz clock frequency, and hence most of the noise generated by this loop is in the filter stop band. The BW controller tunes the resistance of the CMOS resistors through their bias current. For the detection of the filter quality factor, information from the transient response of a biquadratic section is exploited instead of the classical PLL control loops. Hence a non-critical low-frequency 140 kHz clock is employed and high-frequency PLLs and VCOs are avoided. Because in the developed tuning circuits both clock references are in the stop band of the filter, higher dynamic range and lower interference effects can be achieved compared to the other known tuning techniques [12,20,23]. A discussion of the major design considerations for those loops follows.

Automatic tuning of the frequency. For the automatic tuning of the pole frequencies, the charge comparison system (see chapter 4, section 4.4) is employed. For the discussion of the controller, we refer to the charge based controller shown in Fig. 4.8.

The system comprises two weighted dc current sources and a switched capacitor integrator. Signals ϕ_1 and ϕ_2 are two non-overlapping 450 kHz clock phases. In the charge comparator, the unity connected OTA is implemented with the LDOTA2 of the filter, see Fig. 5.17. The common-mode feedback of the LDOTA2 is arranged directly from the LDOTA2 common-mode voltage itself. Hence, for the common-mode stabilization of the LDOTA2, only a simple amplifier is needed. Ideally, the capacitor C_1 is equal to 4.7 pF. This is the average capacitance in the loop of the second biquad. This capacitance determines the frequency of the poles of the second biquad. In the ideal case, the resonant frequency of this biquad is 10.56 MHz.

A clock frequency of 450 kHz has been used. The ratio of the dc current sources is 148. For the design of the current sources, unity and inter-digitized cascode transistors are employed. The bias current of the basic 6μm/6μm transistor is 0.94 μA. The total I_B current is 140 μA, see Fig. 4.8. For this bias current and for $g_{m2} \approx$ 300 μA/V, the differential voltage at the output of the unity connected OTA2 is around 0.93 V. Because the maximum linear range of the OTA is in the order of 1.8 V, then the unity feedback OTA2 can support variations on g_{m2} of the order of 100 %. The integrating capacitor C_H is equal to 6 pF. Hence, the differential jump at the OPAM output voltage is in the range of 0.76 V. However, the single-ended jump is only 0.38 V. The OPAMP is a classical fully-differential folded cascode structure [27]. The gain-bandwidth product of the OPAMP is only 15 MHz and its drain current is 600 μA. It is not necessary to increase this gain-bandwidth product because the clock frequency of the switched-capacitor integrator is only 450 kHz.

In the differential-to-single ended converter, the OTA is a single-ended version of LDOTA2. A LDOTA is used in the converter because: firstly, the large linear range of the low-distortion OTA input-stage increases further the linear operation of the converter and secondly, the common-mode feedback for the OPAMP is arranged from the common-mode voltage implicit in the input-stage of the LDOTA. In addition, this LDOTA allows to adjust the bias current of the transistors M_7 by more than about 100%. It has to be recognized that the linear range of the loop depends strongly of the transconductance of this OTA. For stable operation, the larger the transconductance of this OTA is the lower the dc voltage at the output of the OPAMP is. For the switched-capacitor implemented resistor, a 2.5 pF capacitor has been used. The fixed capacitor of the lowpass switched-capacitor filter is an external, not critical, 100 pF capacitor. However, for on-chip lowpass filter designs a lower capacitor can easily be integrated in the chip.

Automatic tuning of the quality factor. The BW controller has been discussed in the previous chapter, section 4.5. For the discussion of the design considerations, the system shown in Fig. 4.9 is used as the reference. For the implementation of the master filter, 3 OTA2s are employed, see Fig. 4.12. For the reference filter, however, 2 OTA1s and 2 capacitors equal to $2C_1$ are used. The resistor is a replica of the one used in the slave filter. The envelope detectors are implemented with the topology shown in Fig. 4.13, chapter 4 section 4.5. the transistor dimensions and the bias currents of the low-distortion input-stage of the envelope detector are the same as for the OTA2. For M_d and M_{det}, the dimensions, W/L, are 150 μm / 4 μm and 200 μm / 1.5 μm, respectively. The discharging current of the detector, I_{det}, and the holding capacitor of the envelope detector, C_{det}, are 4 μA and 4 pF, respectively. These values lead to a discharge rate of 1 V / μsec.

In the BW controller, the single phase integrator is realized by the OTA, the clock phases ϕ_3 and ϕ_4, the OPAMP and the capacitor C_I. The OTA is a classic single-ended symmetrical OTA. The major design constraint for this OTA is a low dc offset voltage, i.e. lower than 5 mV. The gain-bandwidth product of this OTA is not critical because the low-frequency of the loop. The transconductance of this OTA is 32 μA/V. A higher transconductance should improve the accuracy of the BW system, provided that the dc offset voltage of the topology is not increased. HSpice simulations have shown that this transconductance is large enough for a loop accuracy of 10 %. The

OPAMP is a low dc offset symmetrical OTA. Its gain-bandwidth product and its dc gain are 5 MHz and 55 dB, respectively. C_I is a 10 pF capacitor. For the sample and hold, an external 100 pF capacitor has been employed. The value of this capacitor is non-critical, provided that it can hold the sampled OPAMP output voltage during the holding time. It is also important to point out that the clock feedthrough produces dc offset voltage. For the switches, however, it is very desirable to use minimum transistor dimensions. Finally, the transistors of the level shifter are similar to the transistor used for the biasing of the floating resistors, M_s in table 5.6.

5.4.3. Experimental results. The circuit has been fabricated in a N-well double metal 1.5 μm CMOS process. A microphotograph of the chip is shown in Fig. 5.18. The chip includes the fourth-order bandpass filter, the automatic tuning systems, and several circuits for testing purposes. The chip area is approximately 6 mm^2. The power consumption for the filter together with the automatic tuning systems is around 220 mWatts. The power supplies are only ± 2.5 Volts.

For the characterization of the filter, the setup of Fig. 5.19 has been employed. Hence, the data presented in this section are scaled up by a factor to compensate the attenuation of OTA2 and the 50 Ω resistors.

Frequency response of the filter. The frequency response of the automatically tuned bandpass filter is depicted in Fig. 5.20. For several chips measured, the variations of the center frequency of the filter are ± 100 kHz. These variations correspond to a frequency controller accuracy of ± 1 %. However, the accuracy of the filter bandwidth is around 30 %. The tuning errors will be discussed shortly. The peak gain of the bandpass filter is around -3 dB and the bandwidth is around 300 kHz. The stop-band attenuation at 5 MHz and at 20 MHz are around -60 dB. At very low frequencies, however, the attenuation is in the order of -68 dB. In the experimental results, the noise level of the system makes it extremely difficult to measure very large attenuation's. HSpice simulations have shown dc attenuation's in the range of -85 dB.

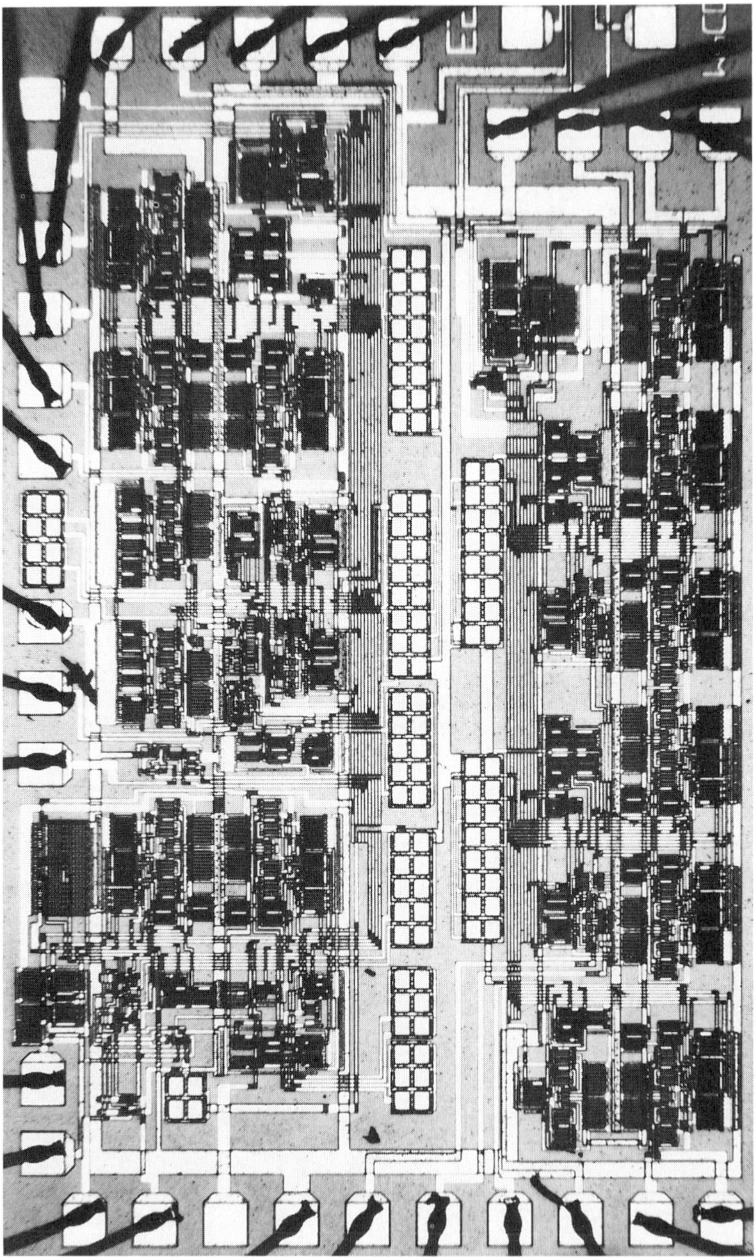

Fig. 5.18. Microphotograph of the chip.

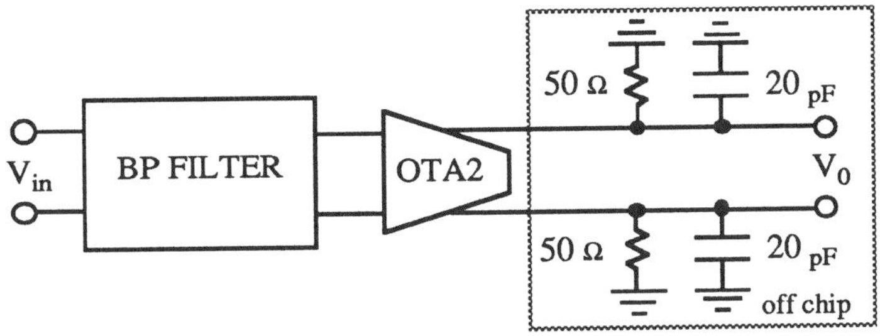

Fig. 5.19. Setup for the bandpass filter characterization.

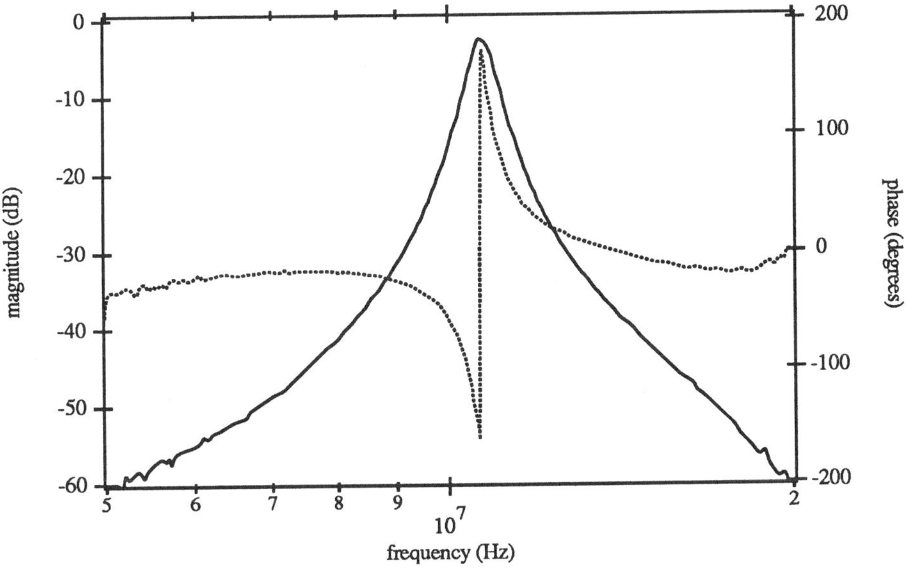

Fig. 5.20. Magnitude and phase response of the automatically tuned bandpass filter.

The very large Q programmability of the filter, manually tuned, is demonstrated in the curves of Fig. 5.21. The filter bandwidth can be controlled from 250 kHz up to 2 MHz. For these results, the bias current of the floating resistors are 10 μA, 50 μA and 250 μA. The quality factor of the filter varies from 5 up to 50. Furthermore, the peak gain of the filter is controlled by more than 30 dB, from -30 dB up to 0 dB.

For the same control currents, the phase response of the filter is depicted in Fig. 5.22. The very large tunability range of the current source based floating resistor is obvious from the results of Figs. 5.21 and 5.22 .

Intermodulation distortion for the filter. The intermodulation distortion for the fourth-order bandpass filter is shown in Fig. 5.23. In this plot, a two-tone signal has been applied to the filter. The frequency of the signals are 10.695 MHz and 10.705 MHz, respectively. The amplitude of the fully differential input signal is 3.2 V_{pp}. In this plot, it can be noted that the third-intermodulation distortion is slightly lower than - 40 dB.

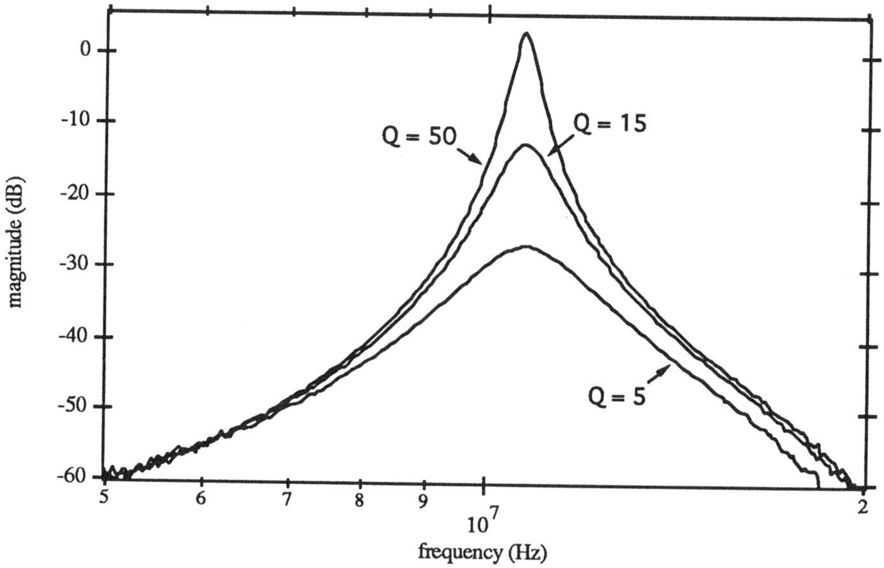

Fig. 5.21. Measured frequency response for several Q factors. the bias current for the floating resistor are 10 μA, 50 μA and 250 μA, respectively.

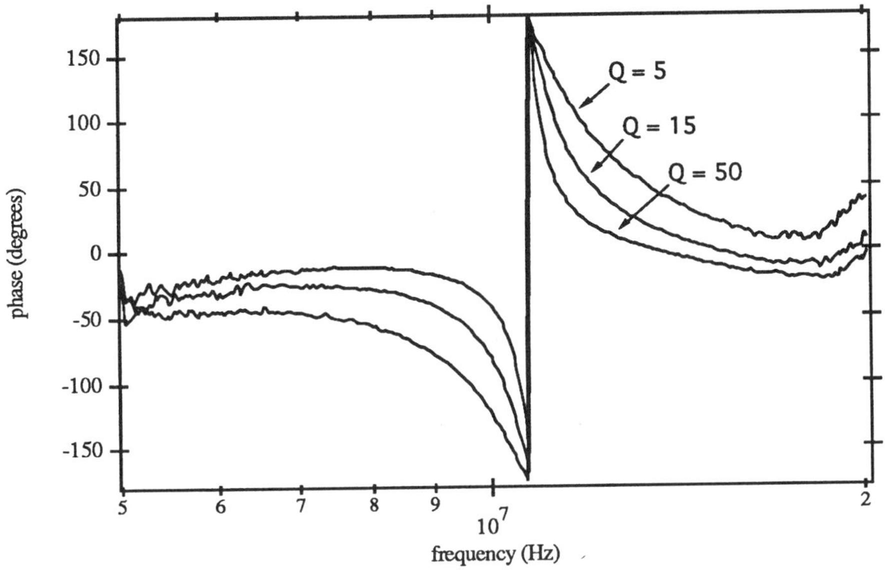

Fig. 5.22. Measured phase response of the filter for several Q factors.

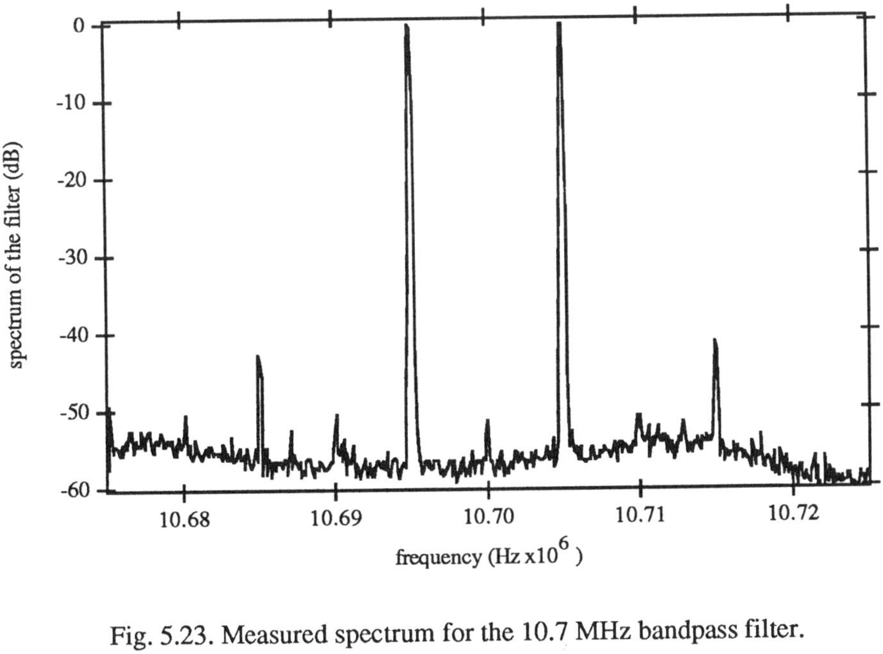

Fig. 5.23. Measured spectrum for the 10.7 MHz bandpass filter.

It can also be noted that the higher-order distortion components, for instance at 10.675 MHz, are more than 5 dB smaller. The harmonic components at the frequencies 10.68 MHz, 10.69 MHz 10.7 MHz, 10.71 MHz and 10.72 MHz are mainly due to the even-order harmonic distortions of the buffer and also due to the higher-order distortions of both LDOTA and resistor. This plot shows a filter linear range for input signals up to 1.2 V_{RMS}.

This spectrum shows the low distortion of the LDOTA but also the low distortion behavior of the active resistor. The reason can be explained as follows. The fourth-order filter is based on two biquadratic sections. The resonant frequency of each section is very close to the center frequency of the filter. For each biquadratic section of the filter, at the resonant frequency its frequency response is dominated by the input transconductor, OTA1 in Fig. 5.17, and the active resistor. The loops, due to the OTA2s and the capacitors, are almost in resonance. Hence, the OTA2s are transferring the charge between the capacitors but they do not interfere with the charge injected by OTA1. The situation is exactly the same as a capacitor in parallel with an inductor and a resistor. At the resonant frequency, the total impedance of the RLC circuit is equal to R. Hence, any injected current with a frequency equal to the resonant frequency of the LC tank is converted to voltage only by the resistor. In the biquad, the equivalent impedance of the capacitors and OTA2s in the loop is very high, in the ideal case infinite. Therefore, the impedance at the output of OTA1 is dominated by the active resistor. As a result, the OTA1 sourced current is injected to the resistor. As a result, the distortion of the filter is dominated by the non-linearities of both OTA1 and active resistor.

Distortion of the OTA2. The distortion for a single OTA2 is shown in Fig. 5.24. This spectrum is for a single output of the fully-differential OTA. The frequency of the input signals are in the pass-band of the filter. For this plot, a 3.6 V_{pp} two tone input signal has been applied to the LDOTA.

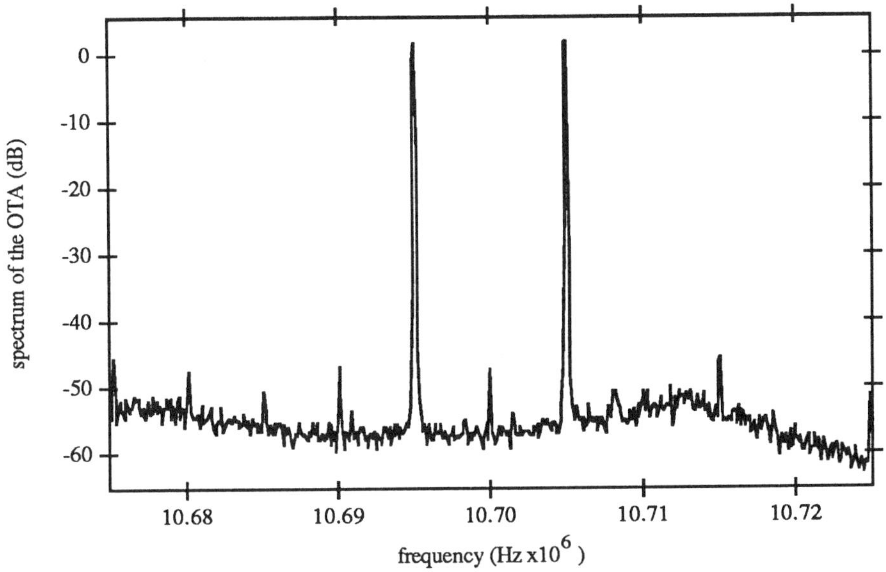

Fig. 5.24. Spectrum for the Low-distortion OTA.

The load impedance of the OTA is a parallel of a 50 Ω resistor and a 20 pF capacitor, as in the case of the filter (see Fig. 5.19). The spectrum of Fig. 5.24 has been scaled up by a factor 50 Ω / 150 μA. Although the single-ended measurement, it can be seen that the intermodulation distortion is relatively low. For this excitation, the third-intermodulation distortion is about -45 dB. The higher order non-linearities of the OTA produces several harmonic distortions, even at 10.7 MHz. Those components appear because the very large input signal applied to the OTA2. It has to be taken into account that the common-mode range of the OTA2 is only 0.8 V. However, for these results the amplitude of the signal at each OTA2 input is 0.9 V.

The filter dynamic range. The filter noise level is dominated by the low transconductance OTAs, i.e. OTA1 in Fig. 5.17. The output referred noise density of the filter is depicted in Fig. 5.25. In this plot, the noise density is shown for the cases when the filter is manually tuned and when it is automatically tuned. It can be noted that the noise density of the filter is not further affected by the automatic tuning systems.

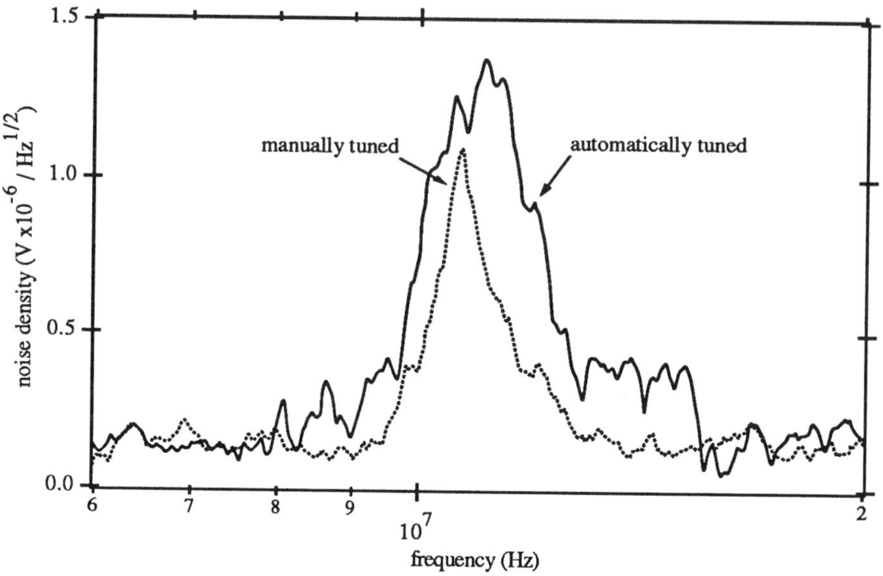

Fig. 5.25. Output noise density of the filter. The noise density of the filter with (line) and without (dash) the automatic tuning system are shown.

In this plot, some spikes can be recognized. These frequency components are mainly due to the BW controller. In the BW controller, the resonant frequency of the biquad is modulated by the applied pulses leading to high-frequency harmonic components. The presence of these harmonic components can be explained from equations 4.28 and 4.29, chapter 4. For the BW controller, Fig. 4.12, the applied pulses to the master filter can be described by the following Fourier series

$$v_P(\omega) = \frac{V_P}{2} + \frac{2\,V_P}{\pi} \sum_{N=1}^{\infty} \frac{\sin\left(N\,\omega_{clock}\,t\right)}{N} \tag{5.4}$$

with V_P the amplitude of the applied pulses. The amplitude of the harmonic component of v_P at the center frequency of the main filter is given by

$$v_P\left(\omega = \omega_0\right) \cong \frac{2\,V_P}{\pi\,N} \cong \frac{2\,V_P}{\pi}\,\frac{f_{clock}}{f_0} \tag{5.5}$$

For this application, however, f_{clock}/f_0 = 140 kHz / 10.7 MHz ≈ 0.014. At the output of the biquadratic section in the BW tuning circuit this harmonic component is amplified by a factor of Q, in this case Q=20. Hence, at the output of the biquadratic section the amplitude of the harmonic component around f_0 is approximately equal to 0.2 V_P ≈ 0.06 Volts. In addition to these harmonic components, other harmonics are present (see equation 5.4). However, most of these components are at the noise level. Due to the unavoidable coupling capacitors, these components are coupled from the BW tuning circuit to the output of the filter leading to the spikes shown in Fig. 5.25.

In theory, the noise feedthrough due to the BW controller is composed only by harmonic components. In Fig. 5.25 these noise contributions should be Dirac deltas. It can be noted in Fig. 5.25 that the amplitude of the spikes are smaller than 1. 5 times the maximum filter noise level. This demonstrate that in fact the noise feedthrough of this automatic tuning system is very small. If the BW controller is disconnected then the noise density of the filter is almost the same as the manually tuned filter. Hence, this demonstrates the low noise feedthrough behavior of the frequency controller.

For the automatically tuned filter, the in-band input referred noise is around 450 μV_{RMS}. The linear range of the filter is about 1.2 V_{RMS}. These values lead to a filter dynamic range of 68 dB, with the IM3 < - 40 dB (1%).

5.4.4. The behavior of the automatic tuning system. For several chips measured, the variations of the center frequency of the automatically tuned filter are around ± 100 kHz. Those results are in good agreement with the ± 1 % accuracy expected from the analysis of the system carried out in chapter 4, section 4.4. The functionality of the frequency controller is shown in the Fig. 5.26. The center frequency of the filter is controlled with the clock frequency of the automatic frequency controller. In this plot, the clock frequency has been varied from 375 kHz up to 525 kHz. The accuracy of the tracking between the measured center frequency of the filter and the clock frequency is around ± 1% over a clock frequency range of 300 kHz up to 600 kHz (center frequencies in the range 7 MHz to 14 MHz). For very high clock frequencies, the major limitation of the accuracy of the system is the offset voltage of the voltage to charge converter (see section 4.4). In the case of low clock frequencies, the accuracy of the frequency controller is limited by the linear range of the switched-capacitor integrator. Nevertheless, the frequency controller works properly for clock frequencies in the range 300 kHz to 600 kHz.

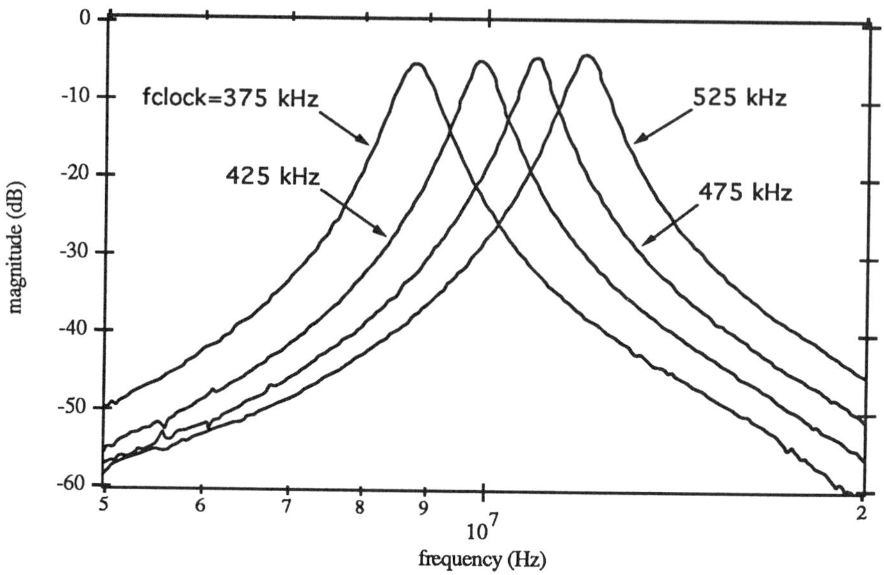

Fig. 5.26. Frequency response of the slave filter for several clock frequencies applied to the frequency controller.

The characterization of the automatic bandwidth controller is rather complicated. The bandwidths of the biquads are fixed by the ratio of the transconductance of OTA2 to the transconductance of OTA1, see Fig. 4.12. Hence, the BW of the biquads can not be programmed, they are fixed. However, the errors in the frequency of the poles affect the bandwidth and the peak gain of the filter, as explained in section 4.5. Hence, it is quite difficult to interpret the errors in the BW of the filter. In the curves shown in Fig. 5.25, the BW errors are as high as + 30 %. However, it is not clear that those errors are completely due to the limitations of the BW controller. It should be remembered that the f_0 sensitivity of the transfer function is proportional to the quality factor of the biquadratic sections. Although the f_0 errors are in the order of 1 %, these errors could produce variations in the BW of the filter in the order of 30 %, see chapter 4 Fig. 4.18. Unfortunately, in higher-order filters, the filter parameters are quite sensitive to the pole frequencies of the biquadratic sections. Hence it is very difficult to separate the BW errors and the errors due to ω_0 variations. A real characterization of the BW controller requires a second-order filter. In a biquadratic filter, the ω_0 errors and the BW errors are not correlated to each other. As a result, they can be characterized independently.

5.4.5. Summary and discussion of the results. A summary of the experimental results is given in table 5.7. As we have already mentioned in the last section, the f_0 and BW accuracy's are in the order of 1 % and 20 to 30 %, respectively. The intermodulation distortion for the filter is lower than 1 % for input signals up to 1.2 V_{RMS}.

parameter	measured
f_0	10.7 MHz
Bandwidth	300 kHz
f_0 variation	≈ ± 1 %
BW variation	≈ 30 %
Vin @ IM3=1%	1.2 V_{RMS}
PSRR (±)	≈ 35 dB
Dynamic range @ IM3=1%	68 dB
Power consumption	220 mW
Power supply	± 2.5
Die size	6 mm^2 *

Table 5.7. Experimental results for the 10.7 MHz filter.
* the chip includes some circuits for testing.

The in-band input referred noise level of the automatically tuned filter is about 450 µV. This noise level leads to a filter dynamic range of around 68 dB, with the IM3 < 1%. All experimental results of the 10.7 MHz filter described here are a consequence of the use of linearized OTAs, linear floating resistors and the low noise feedthrough controllers. Moreover, for this design the supply voltages are only ± 2.5 V. The power consumption of the chip is 220 mWatts and the total chip occupies 6 mm^2. This includes some circuits for test purposes. The filter alone occupies around 1.8 mm^2. The silicon area for both frequency controller and the BW controller are 1.2 mm^2 and 2.2 mm^2, respectively.

In table 5.8, the measured results [24] are compared with other reported realizations [22,23]. The switched-capacitor realization occupies the smallest area but its power dissipation is the larger one. In addition, the dynamic range of the switched-capacitor filter is smaller. The continuous-time realizations, both OTA-C and OTA-R-C, present larger dynamic range. Also, they occupy more silicon area due to the automatic tuning systems needed for continuous-time filters. However, the signal to noise ratio of the OTA-R-C realization outperforms by 17 dB the OTA-C realization. The larger dynamic range of the OTA-R-C realization is due to the linearized LDOTAs used in this topology. This is clear when the linear input voltage range of both topologies are compared. For IM3 < -40 dB, the input signal for the OTA-C realization has to be lower than 0.32 V_{RMS} and for the OTA-R-C realization V_{in} can be as high as 1.2 V_{RMS}.

parameter	SC filter [22]	OTA-C [23]	OTA-R-C [24]
f_0	10.7 MHz	12.5 MHz	10.7 MHz
Bandwidth	250 kHz	250 kHz	300 kHz
Order	6	4	4
PSRR (±)	30 dB	35 dB	35 dB
V_{in} @ IM3=1%	0.24 V_{RMS}*	0.323 V_{RMS}*	1.2 V_{RMS}
Dynamic range @ IM3=1%	34 dB	51 dB*	68 dB
Power consumption	500 mW	360 mW	220 mW
Power supply	10 V	±6V	± 2.5
Technology	2 μm	3 μm	1.5 μm
Die size	2 mm^2	7.8 mm^2	6 mm^2 **

* calculated from reported specifications.

**the chip includes some circuits for testing.

Table 5.8. Comparison of experimental results for several high-frequency filters.

5.5. Conclusions.

In this chapter, the design of several high-performance full CMOS continuous-time filters has been described. In these designs, both ladder filters and biquadratic based filters are employed. The low-distortion and large-dynamic range of the LDOTAs, the CMOS resistors and the filters has been shown. The pole frequencies of the filters range from 100 kHz up to 10.7 MHz. The quality factor of the realizations is in the range of 2 to 50. The versatility and functionality of the OTA-R-C techniques have been shown. Using these techniques, the design of high-frequency high-Q continuous-time filters is feasible.

The low-distortion behavior of the LDOTA has been demonstrated in this chapter. The transconductance of the LDOTA ranges from 6 μA/V up to 300 μA/V. However, the applications of the LDOTA are not restricted to this range. In addition, the experimental results for the fully-differential realizations have shown the feasibility of the technique used for the stabilization of the OTA common-mode voltage. Also, the high linearity and large tunability of the full CMOS resistors predicted in chapter 3 have been pointed out.

The total harmonic distortion for a 100 kHz lowpass ladder filter is lower than -60 dB for two tone input signals up to 2 V_{pp}. Furthermore, the third-intermodulation distortion is lower than - 58 dB for input signals smaller than 1 V_{pp}. For the 10.7 MHz, however, the third-order intermodulation distortion is lower than - 40 dB for two tone input signals up to 3.2 V_{pp}. The dynamic range of the bandpass filter is around 68 dB while IM3 < -40 dB. All those results have been obtained for supply voltages of ± 2.5 Volts.

The functionality of the automatic tuning system has been demonstrated by the control of the 10.7 MHz bandpass filter. The low noise feedthrough behavior of the automatic tuning system proposed in chapter 4 has also been demonstrated. The center frequency of the filter has been controlled with the external clock. The clock controls the filter center frequency in the range of 7 to 14 MHz. The center frequency of the fourth-order bandpass filter is controlled within a ± 1 % accuracy. The accuracy of the filter bandwidth is within 30 %.

References.

[1] A. I. Zverev, *"Handbook of Filter Synthesis"*, John Willey, New York, 1967.

[2] F. E. Girling and E. F. Good, "Active Filters 7 and 8: The Two Integrator Loop," Wriless World, Vol. 76, pp. 117-119, March 1970, and pp. 133-139, April 1970.

[3] F. E. Girling and E. F. Good, "Active Filters 12 and 13: The Leapfrog or Active Ladder Filters," Wireless World, Vol. 76, pp. 354, July 1970, and pp. 445-450, Sept. 1970.

[4] J. Orchard, " Inductorless Filters," IEE Electronics letters, Vol. 2, pp. 155, June 1966.

[5] L.T. Bruton, " Sensitivity Comparison of High-Q Second-Order Active Filter Synthesis Techniques," IEEE Trans. on Circuits and Systems, Vol. CAS-22, pp. 32-38, Jan. 1975.

[6] A. S. Sedra and P. O. Brackett, *"Filter Theory and Design: Active and Passive,"* Matrix publishers, Inc., Champaign Illinois, 1978.

[7] L. P. Huelsman, "*Active RC Filters: Theory and Application*", Editor, Dowden Hutchinson & Ross, Inc., Pensylvania, 1976.

[8] J. B. Hughes, "Self-Tuned RC Active Filters for VLSI," IEE Electronics letters, Vol. 22, pp. 993-994, Sept. 1986.

[9] R. L. Geiger and E. Sanchez-Sinencio, "Active Filter Design Using Operational Transconductance Amplifiers: A Tutorial", IEEE Circuits and Devices Magazine, Vol. 1, pp. 20-32, March 1985.

[10] H. Khorramabadi and P. R. Gray," High Frequency CMOS Continuous-Time Filters," IEEE J. of Solid-State Circuits, Vol. SC-19, pp 939-948, Dec. 1984.

[11] F. Krummenacher and N. Joehl," A 4 MHz CMOS Continuous-Time Filter with On- Chip Automatic Tuning" IEEE J. of Solid-State Circuits", Vol. SC-23, pp 750-757, June 1988.

[12] C. S. Park and R. Schaumann, " Design of a 4-MHz Analog Integrated CMOS Transconductance-C Bandpass Filter," IEEE J. of Solid-State Circuits, Vol. SC-23, pp. 987-996, Aug. 1988.

[13] V. Gopinathan and Y. P. Tsividis," A 5V 7th-Order Elliptic Analog Filter for Digital Video Applications," IEEE Int. Solid-State Circuits Conference, San Francisco, pp 208-209, Feb. 1990.

[14] J. Silva-Martinez, M. Steyaert and W. Sansen, "Design Techniques for High-Performance OTA-R-C Continuous-Time Filters," IEEE J. of Solid-State Circuits, Vol. 27, pp. 993-1001, July 1992.

[15] M. Steyaert, J. Silva-Martinez and W. Sansen, "High Performance OTA-R-C Continuous-Time Filters with Full CMOS Low-Distortion Resistors," IEEE Proc. ESSCIRC-91, Milan Italy, pp. 5 - 8, Sept. 1991.

[16] J. Silva-Martinez, M. Steyaert and W. Sansen," Very Linear CMOS Floating Resistor," IEE Electronics Letters, Vol 26, pp 1611-1612, Sept. 1990.

[[17] J. Silva-Martinez, M. Steyaert and W. Sansen," A Large-Signal Very Low-Distortion Transconductor for High-Frequency Continuous-Time Filters," IEEE J. of Solid-State Circuits, Vol. SC-26, pp 946-955, July 1991.

[18] R. Tarmy and M. S. Ghausi,"Very High-Q Insensitive Active RC Networks," IEEE Trans on Circuit Theory, Vol CT-17, pp. 358-366, Aug. 1970.

[19] J. Silva-Martinez, M. Steyaert and W. Sansen," A high-Frequency Large-signal Very Low-Distortion Transconductor," IEEE Proc. ESSCIRC-90, Grenoble France, pp. 169-172, Sept 1990.

[20] C. F. Chiou and R. Schaumann, "Design Performance of a fully-Integrated Bipolar 10.7 MHz Analog Bandpass Filter," IEEE J. of Solid-State Circuits, Vol. SC-21, pp. 6-14, Feb. 1986.

[21] M. Koyama, H. Tanimoto and S. Mizoguchi, "A 10.7 MHz Continuous-Time Bandpass Filter Bipolar IC," IEEE Proc. CICC-1989, San Diego, pp. 25.2.1 - 25.2.4, May 1989.

[22] Song, B. S., "A 10.7 MHz Switched-Capacitor Bandpass Filter", IEEE J. of Solid-State Circuits, Vol. SC-24, pp. 320-324; April, 1989.

[23] Y. Wang, F. Lu and A. Abidi, "A 12.5 MHz CMOS Continuous Time Bandpass Filter," IEEE Intl. Solid-State Circuits Conference, New York, pp 198-199, Feb. 1989.

[24] M. Steyaert and J. Silva-Martinez, "A 10.7-MHz CMOS OTA-R-C Bandpass Filter with 68 dB Dynamic Range and On-chip Automatic Tuning System," IEEE Intl. Solid-State Circuits Conference, San Francisco, 66-67, Feb. 1992.

[25] M. Steyaert, J. Silva-Martinez and W. Sansen," High Frequency Saturated CMOS Floating Resistor for Fully-Differential Analogue Signal Processors," IEE Electronics Letters, Vol. 27, pp. 1609-1611, Aug. 1991.

[26] J. Silva-Martinez, M. Steyaert and W. Sansen, "A Novel Approach for the Automatic Tuning of Continuous-Time Filters," IEEE Proc. ISCAS-91, Singapore, pp. 1452-1455, June 1991.

[27] P. R. Gray and R. G. Meyer, " *Analysis and Design of Analog Integrated Circuits*," John Willey & Sons, Singapore, 1984.

CHAPTER

6

GENERAL CONCLUSIONS

In this text several topologies and techniques for the design of high-performance full CMOS continuous-time filters have been discussed. Using these topologies and techniques, the design of large-signal (up to 3 Volts peak to peak with power supplies of ± 2.5 V) and low-distortion (total harmonic distortion lower than -60 dB) continuous-time filters is feasible in 1.2-1.5 μm CMOS technologies. Due to the very large linear range of the linearized building blocks, the dynamic range of the continuous-time filters can be increased to the range of 66 - 68 dB. For the tuning of the filters, several on-chip automatic tuning systems are studied. Both the PLL based tuning systems and charge comparison based systems have been described in detail. Using these techniques, continuous-time filters are competitive with the other design approaches, e.g. switched-capacitor filters.

The filters have been implemented using the so-called OTA based technique. The building blocks of these filters are: operational transconductance amplifiers, full CMOS resistors and poly-silicon capacitors. In order to increase the signal to distortion ratio of the filters, low-distortion voltage-to-current transducers have been designed. The odd-order non-linearities are diminished by using these linearized topologies. The even-order distortion components can be reduced using fully-differential and fully-symmetrical structures. In general, a major disadvantage of the fully-differential topologies is the necessity of an additional common-mode control loop. In this text it is shown that using the direct connection of the OTAs, the common-mode feedback can be easily arranged. The result is a very simple common-mode feedback system, hence both power consumption and silicon area are saved. For the control of the filter bandwidth linear CMOS resistors are employed. For low-frequency applications several topologies based on transistors biased in linear region are discussed. For high-frequency and narrow-band applications, two current source based resistors have been described. Compared with the typical OTA based resistors, the major advantages of the CMOS resistors are both their large tunability range and the low distortion sensitivity of the resistance. If the CMOS resistors are used, the versatility and tunability of the continuous-time filters can be further increased. A drawback of this approach is the necessity of a tuning system for the adjustment of the resistance. The whole topologies and the design techniques discussed in the text have been experimentally tested in 3 filter realizations. For all the realizations, the power-supply voltages are only ± 2.5 Volts.

In chapter 2, several OTA schemes have been studied and the discussion results in the design of a low-distortion voltage-to-current transducer. It is shown that the design of efficient voltage-to-current transducers with low supply voltages is feasible. For this topology, its harmonic distortion components can be reduced to the range of - 60 dB for input signals higher than 2 Volts peak to peak. For the voltage-to-current transducer, a current efficiency (the ratio of the dc current to the maximum ac current while the THD < -60 dB) of 42 - 45 % can be achieved. Several OTA practical parameters like dynamic range, frequency response and rejection to supply noise have been discussed. The design strategy for fully-differential OTAs is considered as well. Both the non-linearities and non-idealities of the common-mode feedback have been studied in detail. In this study, the mismatches in both differential path and common-mode loop have been considered. From a simple macro-model which includes all those effects a set of 4 transfer functions is derived. These transfer functions allow the designer to visualize clearly the effects of the OTA non-idealities. Also, the effects of both transistor mismatches and common-mode loop non-idealities on the rejection to supply noise can be easily found out. The effects of the common-mode detector non-linearities are studied as well. Finally, an efficient technique for the design of the CMFB that reduces the complexity of the fully-differential topologies is discussed.

In chapter 3, the design considerations for both continuous-time integrators and linear CMOS resistors are discussed. The effects of the parasitic capacitors on the integrator time constant are studied. The non-linearities of the junction capacitors are considered and their effects on the OTA-C integrator have been pointed out. A simple set of equations relating the filter design constraints with the practical lossless integrator requirements is derived as well. In order to clarify the design procedure for low-distortion continuous-time integrators, two design examples have been included. The first one is devoted to the design of the low-distortion input-stage of the OTA. The second design example is a fully-differential one. The trade-off involved in the design of the common-mode feedback loop is discussed. The experimental results obtained for these designs show the feasibility of these techniques.

For continuous-time filter applications, the major requirements for the resistors are both large tunability range and high linearity. Several CMOS resistors that can be used in the design of lossy integrators are discussed. The advantages and drawbacks of these topologies are pointed out. These topologies implement large resistances with harmonic distortions of the order of -60 dB. In addition, the resistances are almost proportional to the bias current of the resistor and so they can be varied in a wide range, e.g. more than 1 decade. The resistors based on transistors biased in triode region are little sensitive to the transistor mismatches. This is a result of a double linearization procedure for the even-order distortions. For resistances higher than 100 kΩ, the frequency response of this topology is limited to around 1 MHz. The current source based resistors achieve very high-frequency response. This result is due to the reduced number of parasitic poles present in the structure. All these topologies are very well situated for narrow-band applications wherein both small and large time constants are needed in the filter.

In chapter 4, two of the most popular implementations of second-order filters have been discussed. The optimum dynamic range and lower passband sensitivity of the ladder filters make them very attractive for broadband applications, mainly for lowpass filter realizations. For the typical biquad, the dynamic range of the lowpass filter is very bad. Nevertheless, for bandpass filter realizations the biquad presents both higher programmability and good dynamic range. These facts make this type of bandpass filters very attractive for the implementation of narrow band filters. Due to the higher programmability of the filters based on biquadratic sections, these structures can be tuned more accurately than the ladder filters.

The major disadvantage of the continuous-time filters is the necessity of an additional tuning procedure. For on-chip tuning, several techniques have been reviewed. A limitation of several of these techniques is the increment of the main filter noise level due to the tuning signal feedthrough. As a result of the use of the automatic tuning system, the accuracy of the filter is improved but the dynamic range of the automatically tuned filter is reduced. In order to overcome this shortcoming, other approaches for the automatic tuning of the poles frequency have been developed. The charge based technique carries out the automatic tuning of the filter at a different frequency range than the filter passband frequency range. This fact reduces the tuning signal feedthrough at the passband of the main filter. The precision of the frequency tuning loop can be as good as what is obtained with the PLL based techniques, e.g. accuracy's around ± 1%. The unavoidable mismatches make it extremely difficult to improve the precision of the continuous-time filters. For the automatic tuning of the filter bandwidth, a time domain analysis technique has been studied. The information of the filter bandwidth is extracted from the pulse response of a main filter biquadratic section. In this technique a control loop operating at low-frequency is employed. This fact reduces further the tuning signal feedthrough. The design considerations as well as the limitations of the technique are further discussed.

In chapter 5, the feasibility and versatility of the OTA-R-C filters are shown. Both ladder and biquadratic based filters have been designed and characterized. Experimental results for these filters are in good agreement with the theoretical results presented in chapters 2 to 4. The low-distortion behavior of the voltage-to-current transducer developed in chapter 2 is demonstrated by the low distortions obtained for the filters. The common-mode stabilization technique discussed in the second chapter has been used in both a 100 kHz lowpass ladder filter and a 10.7 MHz bandpass filter. The results for those filters have shown the feasibility of the technique. In the realizations, the linear CMOS resistors discussed in chapter 3 are employed. The very large Q tuning of these filters (from 2 up to 50) is a result of the large tuning range of this type of resistors. The CMOS resistors allow to design filters with high-Q and large-signal to noise ratios, even for high-frequency applications.

For a 100 kHz lowpass ladder filter, the total harmonic distortion is lower than -60 dB for input signals up to 2 V_{pp}. The third-intermodulation distortion is lower than - 58 dB for input signals up to 1 V_{pp}. For this distortion, the signal to noise ratio of the filter is approximately 66 dB. For the 10.7 MHz bandpass filter, the third-order intermodulation distortion is lower than - 40 dB for a two tone input signal up to 3.2 V_{pp}. The dynamic range of the automatically tuned 10.7 MHz bandpass filter is around 68 dB, while IM3 < -40 dB. All those results have been obtained for supply voltages of only ± 2.5 Volts.

INDEX